||| || | |||||| || | || | |||||||||||||| || | |||

W9-CAJ-256

TJ   Tabak, John
808.2   Solar and geothermal energy
.T32
2009

                                        MAY   '09

70130

LIBRARY
SOUTHWEST WISCONSIN
TECHNICAL COLLEGE
1800 BRONSON BLVD
FENNIMORE, WI 53809

**ENERGY AND
THE ENVIRONMENT**

Solar and Geothermal Energy

**ENERGY AND
THE ENVIRONMENT**

Solar and
Geothermal
Energy

Knox Learning Center
SW Tech Library Services
1800 Bronson Boulevard
Fennimore, WI 53809

JOHN TABAK, Ph.D.

Facts On File
An imprint of Infobase Publishing

For Archie Shepp, who worked hard to teach me what I wasn't ready to learn. Thank you.

SOLAR AND GEOTHERMAL ENERGY

Copyright © 2009 by John Tabak, Ph.D.

All rights reserved. No part of this book may be reproduced or utilized in any form or by any means, electronic or mechanical, including photocopying, recording, or by any information storage or retrieval systems, without permission in writing from the publisher. For information contact:

Facts On File, Inc.
An imprint of Infobase Publishing
132 West 31st Street
New York NY 10001

Library of Congress Cataloging-in-Publication Data

Tabak, John.
　Solar and geothermal energy / John Tabak.
　　p. cm.—(Energy and the environment)
　Includes bibliographical references and index.
　ISBN-13: 978-0-8160-7086-2
　ISBN-10: 0-8160-7086-5
　1. Renewable energy sources—Juvenile literature. I. Title.
　TJ808.2.T32 2009
　621.47—dc22　　　　　　　　2008026741

Facts On File books are available at special discounts when purchased in bulk quantities for businesses, associations, institutions, or sales promotions. Please call our Special Sales Department in New York at (212) 967-8800 or (800) 322-8755.

You can find Facts On File on the World Wide Web at http://www.factsonfile.com

Text design by Erik Lindstrom
Illustrations by Accurate Art
Photo research by Elizabeth H. Oakes

Printed in the United States of America

Bang Hermitage 10 9 8 7 6 5 4 3 2 1

This book is printed on acid-free paper.

 **Contents**

Preface

Nations around the world already require staggering amounts of energy for use in the transportation, manufacturing, heating and cooling, and electricity sectors, and energy requirements continue to increase as more people adopt more energy-intensive lifestyles. Meeting this ever-growing demand in a way that minimizes environmental disruption is one of the central problems of the 21st century. Proposed solutions are complex and fraught with unintended consequences.

The six-volume Energy and the Environment set is intended to provide an accessible and comprehensive examination of the history, technology, economics, science, and environmental and social implications, including issues of environmental justice, associated with the acquisition of energy and the production of power. Each volume describes one or more sources of energy and the technology needed to convert it to useful working energy. Considerable emphasis is placed on the science on which the technology is based, the limitations of each technology, the environmental implications of its use, questions of availability and cost, and the way that government policies and energy markets interact. All of these issues are essential to understanding energy. Each volume also includes an interview with a prominent person in the field addressed. Interview topics range from the scientific to the highly personal, and reveal additional and sometimes surprising facts and perspectives.

Nuclear Energy discusses the physics and technology of energy production, reactor design, nuclear safety, the relationship between commercial nuclear power and nuclear proliferation, and attempts by the United States to resolve the problem of nuclear waste disposal. It concludes by contrasting the nuclear policies of Germany, the United States, and France. Harold Denton, former director of the Office of Nuclear Reactor Regulation at the U.S. Nuclear Regulatory Commission, is interviewed about the commercial nuclear industry in the United States.

Biofuels describes the main fuels and the methods by which they are produced as well as their uses in the transportation and electricity-production sectors. It also describes the implications of large-scale biofuel use on the environment and on the economy with special consideration given to its effects on the price of food. The small-scale use of biofuels—for example, biofuel use as a form of recycling—are described in some detail, and the volume concludes with a discussion of some of the effects that government policies have had on the development of biofuel markets. This volume contains an interview with economist Dr. Amani Elobeid, a widely respected expert on ethanol, food security, trade policy, and the international sugar markets. She shares her thoughts on ethanol markets and their effects on the price of food.

Coal and Oil describes the history of these sources of energy. The technology of coal and oil—that is, the mining of coal and the drilling for oil as well as the processing of coal and the refining of oil—are discussed in detail, as are the methods by which these primary energy sources are converted into useful working energy. Special attention is given to the environmental effects, both local and global, associated with their use and the relationships that have developed between governments and industries in the coal and oil sectors. The volume contains an interview with Charlene Marshall, member of the West Virginia House of Delegates and vice chair of the Select Committee on Mine Safety, about some of the personal costs of the nation's dependence on coal.

Natural Gas and Hydrogen describes the technology and scale of the infrastructure that have evolved to produce, transport, and consume natural gas. It emphasizes the business of natural gas production and the energy futures markets that have evolved as vehicles for both speculation and risk management. Hydrogen, a fuel that continues to attract a great deal of attention and research, is also described. The book focuses on possible advantages to the adoption of hydrogen as well as the barriers that have so far prevented large-scale fuel-switching. This volume contains an interview with Dr. Ray Boswell of the U.S. Department of Energy's National Energy Technology Laboratory about his work in identifying and characterizing methane hydrate reserves, certainly one of the most promising fields of energy research today.

Wind and Water describes conventional hydropower, now-conventional wind power, and newer technologies (with less certain futures) that are being introduced to harness the power of ocean currents, ocean waves, and the temperature difference between the upper and lower layers of the ocean. The strengths and limitations of each technology are discussed at some length, as are mathematical models that describe the maximum amount of energy that can be harnessed by such devices. This volume contains an interview with Dr. Stan Bull, former associate director for science and technology at the National Renewable Energy Laboratory, in which he shares his views about how scientific research is (or should be) managed, nurtured, and evaluated.

Solar and Geothermal Energy describes two of the least objectionable means by which electricity is generated today. In addition to describing the nature of solar and geothermal energy and the processes by which these sources of energy can be harnessed, it details how they are used in practice to supply electricity to the power markets. In particular, the reader is introduced to the difference between base load and peak power and some of the practical differences between harnessing an intermittent energy source (solar) and a source that can work virtually continuously (geothermal). Each section also contains a discussion of some of the ways that governmental policies have been used to encourage the growth of these sectors of the energy markets. The interview in this volume is with John Farison, director of Process Engineering for Calpine Corporation at the Geysers Geothermal Field, one of the world's largest and most productive geothermal facilities, about some of the challenges of running and maintaining output at the facility.

Energy and the Environment is an accessible and comprehensive introduction to the science, economics, technology, and environmental and societal consequences of large-scale energy production and consumption. Photographs, graphs, and line art accompany the text. While each volume stands alone, the set can also be used as a reference work in a multidisciplinary science curriculum.

Acknowledgments

The author gratefully acknowledges the contributions of John Farison, Director of Process Engineering for Calpine Corporation at the Geysers Geothermal Field, who generously shared his insights and expertise; Elizabeth Oakes, who researched the photos used in this volume; Leela Christian-Tabak, who helped with the statistics; and Frank Darmstadt, executive editor, Facts On File, for his patience and support.

Introduction

Solar energy and geothermal energy have a good deal in common. They are abundant and widely, if unevenly, distributed. They are two of the least environmentally disruptive sources of power available. They are, for the most part, expensive to develop, and compared to more conventional sources of energy, relatively little power is produced from either one.

There is also much to distinguish these two energy sources. Solar energy is intermittent; geothermal energy can be harnessed to produce a continuous stream of power. Solar energy is renewable in the following sense: No matter how much solar energy is converted into heat or electricity at a given location on a given day, there will be more solar energy available for conversion the next day. One cannot ruin a good solar energy site by developing it. Solar energy is inexhaustible. The same cannot be said of geothermal energy. It is possible to destroy even the most productive geothermal site by careless development. This has, in fact, already occurred at some sites.

The subject of solar energy comprises the first half of *Solar and Geothermal Energy*. This section begins by recounting the history of attempts to harness energy from the Sun. The first attempts occurred thousands of years ago, but in the 19th century, efforts to harness the Sun's energy took a surprisingly modern turn due to the efforts of the French engineer and educator Augustin Mouchot. He sought to create heat engines powered by the Sun. The chapter concludes by describing early efforts to convert sunlight directly into electricity via photovoltaic technology.

Chapter 2 describes sunlight in terms of its energy content and availability. Chapters 3 through 5 describe the principal ways that solar energy is harnessed to produce electricity and heat. Chapter 6 describes how solar power plants, which are intermittent power producers, are used to generate power for the grid, and chapter 7 describes some of the ways that the United States and Germany have encouraged the development of the solar power industry. Government policies are important because at its present level of development, the solar industry would collapse without substantial government support.

The second half of the book is concerned with geothermal energy. Chapter 8 describes how geothermal heat sources were first discovered and how they are distributed about the globe. It concludes with a brief history of early efforts to convert thermal energy from Earth's interior into electricity. Chapter 9 provides an overview of geothermal technology, and chapter 10 describes some of the details of geothermal heat engine design, and contains an interview with John Farison, director of process engineering for Calpine Corporation at the Geysers Geothermal Field, about maintaining power production at the Geysers. Chapter 11 describes so-called *direct use* technology, which uses thermal energy from beneath Earth's surface without converting it into electricity. Chapter 12 describes some of the economic aspects of geothermal energy, and chapter 13 describes some of the ways that various agencies, national and international, are working to develop the technology needed to expand the use of geothermal energy in the power-generation sector.

A strange relationship exists between the amount of energy available from a particular source and the extent to which that source is used: The greater the amount available, the less it is utilized. In particular, solar energy and geothermal energy are two of the most abundant sources of energy available, and yet they are two of the least utilized. For those interested in the environmentally responsible production of electric power, it is important to understand the potential of these energy sources and why, so far, that potential has yet to be realized.

PART I

Solar Energy

A Brief History
of Solar Power

The impact of solar technology on society has, so far, been small, with one major exception—the use of solar panels to power spacecraft. In the United States, for example, less than one-tenth of 1 percent of the nation's electricity supply comes from solar generating stations.

But if the contribution made by solar energy technologies has been modest compared with many other more conventional energy sources, the same cannot be said for claims about its potential. For decades, advocates of solar power have downplayed the difficulties involved in harnessing the Sun's energy, preferring instead to predict quick and widespread adoption of various solar technologies. A great deal of 1970s literature, for example, confidently predicted that by the year 2000 the United States would have a solar-powered economy in which solar arrays and solar water heaters would be everywhere, and dependence on oil would be much reduced. This, of course, has yet to happen. There is little evidence of an imminent solar revolution. But

A fanciful depiction of Archimedes using a mirror in the defense of Syracuse. Thousands of years ago, the Greeks knew that mirrors could be used to concentrate sunlight.

there is no denying the potential of solar energy to make an important contribution. The Sun is inexhaustible, and converting its energy into heat and electricity are relatively clean processes.

This chapter provides an overview of the history of solar power technologies. These technologies fall into one of two categories: *photovoltaic* (PV) technologies, which convert sunlight directly into electricity, and those technologies that convert sunlight into thermal energy. Both are important. Thermal technologies, which are older, are described first.

SUNLIGHT AND HEAT

Passive solar is the term used to describe a suite of technologies that use the Sun's energy directly, usually for heating and cooling, and without the aid of any mechanical or electrical devices. For example, many technologically simple cultures built homes and other structures so that the interiors of the buildings could absorb the maximum amount of sunlight during the coldest days of the year. Many ancient Indian and Greek homes, for example, were built with this in mind. The Sun shone into the buildings and heated thick walls and floors during the cooler months, and by using overhangs, the same interiors could be shaded from the summer Sun, which is higher in the sky during the middle of the day. These early solar designs would probably have seemed almost obvious to those who depended upon them, an assertion supported by the fact that some of these ideas were used in cultures widely separated by time and distance. The reason is clear: When compared to modern societies, early societies were energy-poor, and the Sun's energy would have made a welcome contribution to their energy supply.

The Sun's energy is diffuse, and at Earth's surface it is also intermittent and unreliable. Consequently, when fossil fuel–fired technologies were developed that allowed users to produce large amounts of heat whenever and wherever they wanted, people everywhere adopted them. Homes, factories, and public buildings were constructed with little regard for the Sun as a source of heat. Interest in the Sun as an energy source waned as per-person energy usage soared. This revolution in energy usage began with coal; next was oil, and other energy sources soon followed. Some designers continued to build with the Sun in mind, but only where and when the price of fossil fuels was high.

Following the 1970s energy crises, interest in passive solar design again increased but only briefly. Interest diminished following the collapse of oil prices during the 1980s, but increased during the 1990s. Today, in some areas, most notably the Netherlands and Germany, new building designs incorporating passive heating and cooling have become fairly common. These designs tend to be more expensive, because they are designed to make use of local conditions such as local solar intensity and the orientation of the building relative to the Sun's path across the sky. This serves to illustrate the fact that with respect to passive systems, no single design works everywhere. But many energy-saving design features that draw interest today—features such as using the Sun to heat a very massive floor which will then radiate heat throughout the cooler evening, or using sunlight to power convection currents that circulate air throughout all or part of a building—have once again become part of some architects' repertoire.

Passive systems are valuable, but technologies that use sunlight to drive machinery have more far-reaching implications for society and the environment. These technologies convert sunlight into work, and by far the most common method of converting sunlight into work uses a two-step process. First, light is converted into heat; second, the heat is converted into work. Devices that convert heat into work are called heat engines. All of the big solar-powered electric-generating stations in operation today are heat engines, but heat engines have many uses in addition to electric power generation. Broadly speaking, heat engines, no matter their source of heat, are designed to turn a shaft. In the case of solar-powered generating stations, the shaft is connected to a generator, which is the component of the power plant that actually produces electricity. But a spinning shaft can just as easily be attached to a water pump, or the drivetrain of a car, or a multitude of other devices. Heat engines currently produce most of the power used by society. Coal-fired steam engines, the oldest type of heat engine technology, have been in use for several centuries. Less widely appreciated is that solar-powered heat engines of remarkably modern design were first constructed in the 1860s and served as a foundation for a viable solar industry that lasted about 40 years.

Solar-powered heat engines have two main components—a device for converting sunlight into heat and an engine that converts the heat into work. Devices capable of converting light into heat have been known since antiquity. The use of lenses to concentrate the Sun's rays was described in the play *The Clouds* by the Greek playwright Aristophanes (ca. 450–388 B.C.E.). In the following lines of dialogue between the characters Strepsiades and Socrates, Strepsiades describes how he used a crystal lens to focus the rays of the Sun in order to melt wax.

> Strepsiades: Have you ever seen a beautiful transparent stone at the druggist's with which you may kindle a fire?
>
> Socrates: You mean a crystal lens.
>
> Strepsiades: That's right. Well, now if I placed myself with this stone in the Sun and a long way off from the clerk, while he was writing out the conviction, I could make all the wax, upon which the words were written, melt.

Evidently, the properties of lenses were familiar to many even 24 centuries ago.

Mirrors have also been used to focus the Sun's rays for millennia. *On Burning Mirrors* is a treatise by the Greek mathematician Diocles (ca. 240–180 B.C.E.) in which he describes the optical properties of mirrors and, in particular, the way that a parabolic mirror focuses light to create heat. This work, together with the story of how, under the direction of Archimedes (ca. 290–212 B.C.E.), Greek soldiers at Syracuse focused sunlight on Roman ships as they approached the city's seawall and set them on fire, inspired generations of scholars to study optics in general and burning mirrors in particular. Of special note is the Arabic mathematician and physicist Abu Ali al-Haytham (ca. 965–1039 C.E.) and the German-born priest and scholar Athanasius Kircher (1601–80), both of whom made important discoveries in optics and in the use of burning mirrors.

But there is more to a heat engine than a source of heat. The first machines capable of converting heat into work were steam engines, and the first steam engines were invented by the English inventors Thomas Savery (1650–1715) and Thomas Newcomen (1663–1729).

Their engines, especially Newcomen's, were used for decades without design changes to drive the pumps that cleared coal mines of water. Steam engines were radically redesigned by the Scottish inventor James Watt (1736–1819) and his associates. As the principles by which the engines operated became better understood, inventors developed a great many creative designs, and the new engines were used in new and important ways. By the time of Watt's death, the steam engine powered many industrial processes as well as trains and steamboats.

It is a remarkable fact that by the 1860s, less than 50 years after the death of Watt, steam engines were being built to operate on solar power. The first solar-powered steam engines were designed and built by the French inventor and mathematician Augustin Mouchot (1825–1912). Mouchot believed that coal, which at the time was the only fossil fuel in general use, would soon be exhausted. Many of his contemporaries shared his belief. Mouchot believed that solar energy was a viable substitute, and he devoted his efforts to creating practical solar-powered heat engines. Mouchot began by building devices for concentrating solar energy and using them to make solar ovens and devices for distilling alcohol. By 1866, he was running a steam engine on solar energy, and by 1878, he was successfully powering an ice maker with sunlight.

The French government was especially interested in using solar engines in what was then the French colony of Algeria, a location where coal was expensive and power was in short supply. Mouchot received substantial government support for his efforts. Recognizing that the intermittency of solar energy was a serious problem, Mouchot sought to design solar-powered steam engines that could operate continuously, just as coal-fired steam engines can, in theory, operate for as long as fuel is available. The key, in Mouchot's view, was to convert solar energy to a form in which it could be easily stored. He considered using his collector to heat water, which could then be stored in large insulated tanks. Later, when heat was needed, the hot water could be withdrawn and used to power a heat engine. (A power plant using this concept, called Solar Two, was built in California more than a century later, but it used molten salt rather than water to store the thermal energy.) In the end, however, Mouchot decided to use the Sun's energy to create an electric current. The current was used to reduce water into its elemental constituents, hydrogen and oxygen. Hydrogen, which is highly flammable, could be stored and burned later when energy was needed and the Sun was not available. (This is another technique for storing solar energy that is often discussed today.) Although contemporary solar-powered heat engines benefit from better materials and better

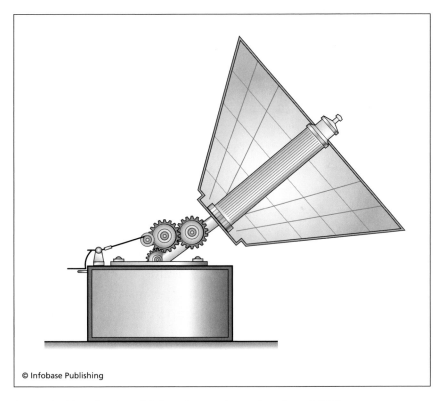

© Infobase Publishing

Augustin Mouchot's multiple-tube sun-heat absorber of 1878

manufacturing techniques, many of their most important design concepts can be found in the work of Mouchot.

Influenced by Mouchot's work, a number of inventors, some already very distinguished, sought to generate profits from the Sun. A notable example is the American inventor Frank Shuman (1862–1918). Already wealthy from patents he obtained for safety glass and wire-reinforced glass, Shuman applied himself to building solar-powered water pumps for large irrigation projects. Because such pumps do not need to operate continuously, the problem of storing solar energy did not need to be addressed, and the economics seemed promising since demand was greatest where coal, the principal fuel source at the time, was most expensive. One pump, a demonstration model built in Egypt, produced 65 horsepower (48 kW) and pumped 6,000 gallons (23,000 l) of water per minute. It was designed specifically to show how the Sun could be used to irrigate the Sahara, but the broader implications of solar energy seemed clear to many. Some engineers of the time predicted that Egypt would become an important industrial center because of its enormous solar resources, an idea that has begun to attract attention again today. (DESERTEC, an organization supported by Prince Hassan of Jordan, is presently evaluating the potential of huge solar-powered thermal generating units in the deserts of the Mideast and North Africa and exporting much of that power to Europe.)

The work of Mouchot, Shuman, and others from this highly creative period in the history of solar power ended with the onset of World War I. During that war, oil's value as an energy source became apparent to all. It was relatively easy to transport; it was energy-rich, and it could power a wide variety of devices. And enormous deposits of oil had been discovered in the United States and several other places. So much oil was produced that in some of the bigger fields it was sometimes sold for pennies per barrel. Solar energy could not compete. Solar-powered engines, although their fuel was free, were expensive to build and were ill-suited to cloudy envi-

The Photophone: A Solar-Powered Telephone

In 1880, Scottish-born American inventor and educator of the deaf Alexander Graham Bell (1847–1922) and his research assistant, Charles Sumner Tainter, successfully tested a device they called the photophone. Already a holder of numerous patents, the most famous of which was for the telephone, Bell believed the photophone to be one of his greatest inventions. So jealous was he of the device that shortly after its invention, he and Tainter deposited sealed boxes containing a model of the photophone together with accompanying documentation at the Smithsonian Institution to protect their claim to priority. (The boxes were not opened until 15 years after Bell's death, long after the technology had become obsolete.) Bell's device encoded the sound of voices on rays of sunlight and enabled the user to send a message from one point to another without wires.

The key to the photophone was the use of the metal selenium in the device's circuits. When light shines on selenium, the metal's electrical properties change almost instantaneously.

The photophone used a mirror to direct sunlight through a lens. The lens focused light, which then shone through a mechanical device that vibrated to the sound of a human voice. As the device vibrated it varied the intensity of the light passing through it. (To see how this worked, imagine shining a light on two combs, one aligned behind the other and both pointing in the same direction. If one comb moves relative to the other, sometimes the teeth of the two combs will be aligned and sometimes they will be misaligned relative to each other. When the teeth are aligned more of the light will pass through them. When the teeth are misaligned, they will block a larger percentage of light.) Bell had found a way to modulate the intensity of a beam of sunlight so as to carry information about a human voice.

Once past the modulator, a mirror was used to reflect the flickering light toward an electrical circuit containing selenium. As the electrical properties of the selenium changed in response to changes in the light's intensity, the electrical current that passed through the circuit varied

(continues)

(continued)

in intensity as well. Bell used this variable electrical current to drive the speaker in a telephone earpiece. In this way, Bell was able to send a voice on a beam of sunlight.

Bell was extremely proud of his invention and believed that it would eventually have important applications. The Sun, he believed, would power communications. But unlike some of his other inventions, the photophone was never more than a curiosity. In Washington, D.C., where Bell first sent a message with his photophone, the Sun is unreliable as a source of direct light. Clouds are frequently in the sky. (In letters to his wife, Mabel Hubbard Bell, he describes his frustration when an experiment would be cut short by a passing cloud.) In addition to clouds, the Sun's light can be blocked by smoke, dust, and many other randomly occurring phenomena. As a consequence, the photophone was always unreliable.

Although impractical as a communications device, Bell and Tainter's invention indicates that even in 1880, it was possible for those with enough imagination to harness the Sun's energy to do useful work—in this case, to drive a speaker. Bell's work further illustrates the difference between theory and practice in the utilization of solar energy, a difference that is just as important today as it was more than a century ago.

ronments, and despite Mouchot's efforts to build plants that worked continuously, all solar engines stopped at sunset. Research efforts into solar-powered heat engines did not begin again in earnest until after the energy crisis of 1973.

EARLY PHOTOVOLTAIC TECHNOLOGY

Photovoltaic (PV) technology converts sunlight into electricity without first converting it into heat. The history of PV technology apparently began when Alexandre-Edmond Becquerel (1820–91), working in the laboratory of his father, the French physicist An-toine-César Becquerel (1788–1878), observed that he could generate an electric current by exposing certain materials to light. (Some say the father was the first to observe the effect.) But the phenom-enon that both Becquerels observed was weak and poorly under-stood, and only modest progress was made in understanding the relationship between light and electricity during the 19th century. The most significant accomplishment to occur after the Becquerels' experiments and prior to 1900 occurred in 1883, when American inventor Charles Fritts created the first true solar cell. Fritts used selenium, the same material used by Alexander Graham Bell in the construction of his photophone. (See the sidebar "The Photophone: A Solar-Powered Telephone.") His accomplishment was described in "On a New Form of Selenium Cell" and published in 1883 in *The American Journal of Science*. Fritts's cells were an interesting and important technical accomplishment. He understood the potential for the technology, but from a practical point of view his cells were useless because they converted less than 1 percent of the incident light into electricity.

An important theoretical breakthrough occurred in 1905 with the publication of a research paper by German-born physicist Albert Einstein, in which he partially explained the photoelectric effect, the phenomenon that is responsible for the generation of electricity by a solar cell. For this paper, Einstein would eventually receive the Nobel Prize in physics. During the next few decades, experimental work was performed regarding the photoelectric effect, but it was not until 1954 that the first photovoltaic cells capable of producing significant amounts of electricity were produced. These cells, which used silicon and operated at 6 percent efficiency, were a product of Bell Labs.

The discoveries of the early 1950s occurred just prior to the launch of the first artificial satellites, a time when the United States and the Soviet Union competed fiercely to launch progressively more sophisticated technology into space. The race began on October 4, 1957, when the Soviet Union launched *Sputnik 1,* the world's first artificial satellite. But this satellite depended exclusively on batteries, which placed severe limitations on the useful life of the satellite. Additionally, batteries are heavy, which reduced the amount of useful payload that could be carried aloft. A better power source was needed.

Soviet- and American-made solar cells from the early days of the space race were expensive and inefficient, but there was no realistic alternative. PV technology provided a continuous and long-term source of power. In 1958, the United States launched *Vanguard 1,* which was outfitted with a PV array that produced somewhat less than one *watt* of power. The Soviet Union launched later versions of Sputnik with PV technology as well. Just as PV technology powered the electronics for much of the hardware sent into space by the United States and the Soviet Union, the pressures of the space race motivated additional research into PV technology.

Progress in PV technology for space applications was steady and rapid. In 1964, only six years after the launch of *Vanguard 1,* the National Aeronautics and Space Administration (NASA) launched the first of seven Nimbus satellites. Designed for research in the atmospheric sciences, the first Nimbus had a PV array capable of generating 470 watts. Today, each PV panel on the *International Space Station* produces approximately 33 kilowatts (33,000 watts) of power.

But if PV technology has proven essential for satellites in Earth orbit and for missions in the inner solar system, it has so far been less valuable for applications on Earth—but not for want of trying. Researchers continue to obtain substantial improvements in PV performance as well as reductions in cost. Cells that convert in excess of 40 percent of sunlight into electricity have been successfully tested, and less efficient PV cells have been produced that are much cheaper to manufacture in bulk than past designs. Despite these advances, PV technology has only been deployed in niche applications. Devices located far from grid connections have benefited from PV arrays, and some backup battery systems, for example, are now charged with PV arrays. And demonstration projects abound, provided they receive substantial government subsidies. The subsidies are needed because despite improvements in price and efficiency, PV-produced electricity remains the most expensive electricity available, which is why it often makes sense to choose some other power source.

In addition to high costs for the arrays themselves, PV penetration into the power sector continues to be limited by the absence of efficient and cost-effective ways of storing electricity for later use. In small applications such as single-family homes, some PV-generated power can be stored in batteries, but batteries are expensive, and so is the hardware needed to regulate the electricity that flows to and from the batteries. But without a mechanism for storing power, the intermittent nature of sunshine means that PV arrays cannot function as stand-alone units. Consequently, home owners interested in solar power must pay for both the PV array as well as a backup system, and two systems are always more expensive than one.

For large-scale uses, only a small number of PV demonstration plants have been constructed. These are expensive to build, useless at night, and unreliable during the day since a single cloud can cause a significant drop in power. Because cloudless days cannot be predicted far in advance, these units require significant and readily available backup. Evidently, PV technology still must overcome significant challenges before it becomes economically attractive. But these problems do not appear to be fundamentally unsolvable. PV technology may yet make an important contribution to the supply of electricity for the grid.

Research continues. In 2007, the Defense Advanced Research Projects Agency (DARPA) announced a three-year effort to build solar cells capable of converting in excess of 50 percent of incident sunlight into electricity.

Sunlight and Geometry

Solar energy is abundant and inexhaustible, but in the United States less than one-tenth of 1 percent of the nation's electricity supply is obtained from solar energy. In all other large economies, whether they are located in regions of abundant direct sunlight or in regions with intermittent direct sunlight, the situation is similar: Solar power makes a very small contribution to the energy mix. To understand the challenges involved in using solar energy, it is necessary to understand the resource in some detail. This chapter begins with a description of the characteristics of sunlight.

When sunlight is intercepted on Earth's surface (rather than in space), the characteristics of Earth's surface and its atmosphere become extremely important. Earth's surface is curved, and Earth rotates rapidly. Consequently, the amount of sunlight striking a flat plate placed parallel to the surface varies enormously from one location to the next and from one hour to the next. Understanding

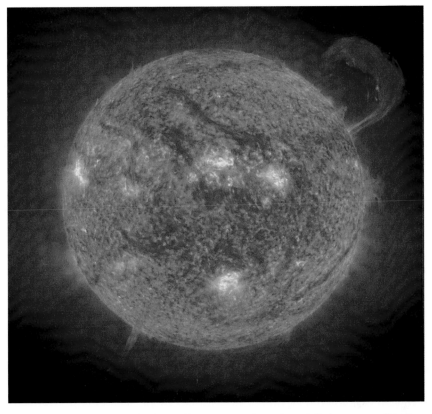

The Sun in ultraviolet light. Different wavelengths of light carry differing amounts of energy to Earth's surface. *(NASA)*

local conditions contributes to an appreciation of why solar energy has been so difficult to harness. This chapter concludes with a description of some of the ways that the geometry of Earth and the properties of Earth's atmosphere complicate the harnessing of solar energy.

THE ELECTROMAGNETIC SPECTRUM

Nuclear fusion is the source of the Sun's energy. In the interior of the Sun, less massive hydrogen atoms are fused together to form more massive helium atoms, and in the process enormous quantities of energy are released. The Sun has been emitting energy in this way for several billion years, and theory predicts that the process will continue for billions of years into the future. The Sun's energy output is remarkably consistent, varying by no more than one-tenth of 1 percent.

The Sun emits an enormous quantity of energy, only a tiny part of which is intercepted by Earth. The energy propagates through space in the form of *electromagnetic waves,* which are periodic variations in the intensity of electric and magnetic energy. These periodic variations can be visualized as sinusoidal, water-like waves, in which case the *wavelength* of an electromagnetic wave would be the distance from the crest of one wave to the crest of an adjacent wave. Light waves and radio waves are probably the most familiar examples of electromagnetic waves. Electromagnetic waves are distinguished principally by their wavelengths. Radio waves have wavelengths that are fairly long; they vary from meters to millimeters in length. *Infrared* radiation, which is perceived as heat, is characterized by electromagnetic wavelengths that vary from about one-tenth of a millimeter to about 700 nm, where the abbreviation nm stands for *nanometer,* or one-billionth of a meter. Visible light consists of electromagnetic radiation with wavelengths in the range 700–400 nm. Different wavelengths within this interval are perceived as different colors, and the entire spectrum of visible electromagnetic radiation can be seen in a rainbow, where the longer wavelengths appear toward the outer rim of the rainbow. Red is the color with the longest wavelengths. The shorter wavelengths appear near the inside rim of the rainbow. Violet is the color with the shortest wavelengths. Electromagnetic waves with wavelengths much shorter than about 400 nm or much longer than 700 nm cannot be perceived by human eyes. *Ultraviolet waves,* which are responsible for sunburns, have wavelengths that are shorter than violet's.

The Sun's energy radiates out in all directions, and as it spreads out, it becomes weaker. But even after crossing the great expanse of empty space that separates the Sun from Earth, a flat plate placed

Knox Learning Center
SW Tech Library Services
1800 Bronson Boulevard
Fennimore, WI 53809

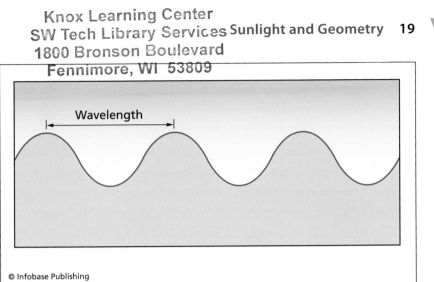

© Infobase Publishing

Wavelength is the distance from one peak to the next (or one trough to next) in a regular wave form.

above the atmosphere and oriented perpendicularly to the Sun's rays would receive the Sun's electromagnetic energy at a rate of about 1,368 watts per square meter, a quantity called the *solar constant*. (The solar constant is always described in terms of watts per square meter—never in imperial units.) Despite its name, the solar constant is not constant. The main source of variation in the solar constant—it only amounts to about 3 percent of its value—is caused by small changes in the Earth-Sun distance, and this variation is due to the slightly elliptical path about which Earth orbits the Sun.

The solar constant represents the "raw material" with which solar engineers have to work. Solar cells convert solar energy into electrical energy, and consequently, the output from photovoltaic cells cannot be greater than the input they receive from the Sun. No solar energy device located on Earth or in Earth orbit can provide more than 1,368 watts per square meter. In practice, however, even in Earth orbit, the output of a solar energy conversion device is much less than the solar constant, and for terrestrial applications, the output of every solar energy conversion device is, when averaged over a day, always a small fraction of the solar constant. (The reasons for thus disparity are described later in the chapter.)

The collection of electromagnetic waves emitted by the Sun carries the Sun's energy to Earth, but different parts of the electromagnetic spectrum carry differing amounts of energy. The Sun emits energy most intensely at about 470 nm, which is in the visible part of the spectrum. Electromagnetic rays with wavelengths at or near 470 nm lie in the yellow-green region of the spectrum. The amount of energy emitted by the Sun drops off rapidly as the wavelengths of the electromagnetic rays diverge from 470 nm. In fact, from an energy point of view, the electromagnetic waves emitted by the Sun are often divided into three categories: electromagnetic waves with wavelengths shorter than 400 nm, electromagnetic waves with wavelengths longer than 400 nm but shorter than 700 nm (this, again, is most of the visible portion of the spectrum), and electromagnetic waves with wavelengths longer than 700 nm. The Sun emits a little more than 40 percent of its energy at visible wavelengths, which occupy a very narrow band in the electromagnetic spectrum. The Sun emits about 50 percent of its energy (but at a much lower intensity) across the very wide part of the spectrum consisting of electromagnetic rays with wavelengths longer than 700 nm. What little energy remains is carried by electromagnetic waves with wavelengths shorter than about 400 nm. Almost all of the Sun's energy is carried by electromagnetic radiation with wavelengths between 200 nm and 4,000 nm. (The value 4,000 nm is usually written as four micrometers and is abbreviated 4 μm, where one micrometer is defined as one-millionth of a meter.)

These basic facts about the intensity of the Sun's rays and the way that energy is distributed among different parts of the solar spectrum are important for designers of the solar panels used on telecommunications satellites and the *International Space Station* (ISS), but by themselves they are of little help for engineers interested in designing terrestrial solar-powered devices.

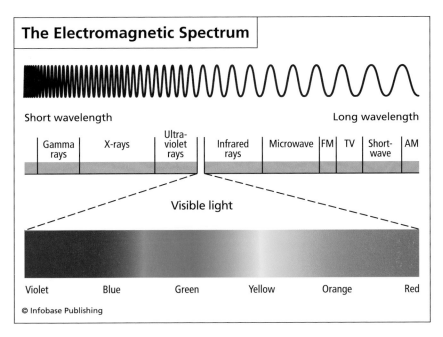

Electromagnetic spectrum. Electromagnetic waves are classified according to their wavelengths. Visible light occupies only a small segment of the spectrum.

Earth's atmosphere filters incoming solar radiation and changes the composition of sunlight. The ozone layer, for example, a region of atmosphere between six and 30 miles (10–50 km) high, prevents many ultraviolet rays from reaching Earth's surface. (Ozone is a molecule formed by three oxygen molecules, and while ozone can be found in the air near Earth's surface, it is concentrated primarily in the region that bears its name.) Most ultraviolet rays with wavelengths less than about 290 nm are absorbed in the ozone layer. The energy carried from the Sun to Earth by these ultraviolet rays is, therefore, unavailable for use at Earth's surface.

The role of clouds is more complex. Low, thick clouds, called stratocumulus clouds, reflect some of the incoming light back into space. This reflected light is, of course, unavailable for use by solar energy devices located on the surface. But high, thin clouds, called

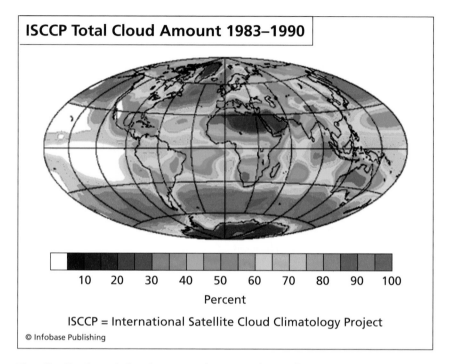

ISCCP Total Cloud Amount 1983–1990

10   20   30   40   50   60   70   80   90   100

Percent

ISCCP = International Satellite Cloud Climatology Project

© Infobase Publishing

The distribution of clouds is complex, as is their effect on the availability of solar energy. *(Source: NASA)*

cirrus clouds, transmit most of the incoming light. Clouds introduce an element of randomness into the production of solar power. It is often difficult to predict far in advance what types of clouds (if any) will be present at a particular location or even whether clouds will be present at all. This element of randomness is important because clouds degrade the performance of most types of solar energy devices and contribute to what is called the intermittence of solar energy. In space, solar energy is never intermittent. On Earth, intermittence is a key characteristic of solar energy.

Other phenomena also degrade the electromagnetic waves that pass through the atmosphere. Airborne sand and dust, ash emitted from volcanoes and forest fires, and particulates emitted as a result of the combustion of fossil fuels, for example, all degrade the ability of the atmosphere to transmit the Sun's energy to the surface. Water vapor also absorbs some short-wavelength energy and emits it as longer wavelength energy. This list of phenomena is a long one, and their relative importance in degrading sunlight depends on location, the time of year, and factors that, like cloud cover, are often difficult or impossible to predict in advance. Under the best of conditions, only about 70 percent of the Sun's light reaches Earth's surface. This amounts to a maximum available input for a solar device of about 1,000 watts per square meter.

But the condition of the atmosphere is just one factor in determining the amount of solar energy that can be converted by any solar energy device. Another set of factors, which are geometric in nature, are even more important.

GEOMETRY AND SUNLIGHT

In order to appreciate the potential of solar energy, it is necessary to understand how Earth's geometry affects the amount of solar energy incident at each location on Earth. Imagine a thin flat disk with a radius equal to that of the Earth, and imagine positioning that disk along Earth's orbital path. Finally, imagine orienting the disk so that the Sun's rays are perpendicular to its face. Such a disk would receive the same amount of solar energy as Earth does. But on a disk, that energy would be evenly distributed across its surface. On a sphere, the situation is very different because on a sphere only a tiny portion of the surface is ever perpendicular to the Sun's rays at any given time. Everywhere else, the sphere's surface slopes away from the Sun's rays, and as a consequence those rays strike the sphere's surface at oblique angles or not at all.

The angle at which the Sun's rays strike a location on Earth's surface is critical when attempting to convert solar energy into electricity or heat. It is a common experience that the higher in the sky the Sun is, the more intense the sunshine. If the Sun is directly overhead—that is, if the Sun's rays are perpendicular to a piece of

⏻ Space-Based Solar Power

On Earth the amount of power received by the Sun fluctuates from zero at night to a maximum of about 1,000 watts per square meter (the exact amount depends, in part, on location, season, and weather). Averaged over a 24-hour period, however, the amount of energy reaching the surface is never more than a few hundred watts per square meter, despite the fact that just above Earth's atmosphere, the Sun delivers an almost constant power of 1,368 watts per square meter. Clearly, when choosing a location for a solar panel, it makes a huge difference whether the panel is on Earth or in space. In 1968, the Czech-born American engineer Peter C. Glaser proposed building a satellite that would convert solar energy to microwave energy and beam that energy back to Earth to be converted into electricity for use on the grid. (Glaser, who was then with NASA, was the project manager for the Laser Ranging Retro-Reflector Array that *Apollo 11* installed on the Moon, the only Apollo project still in use.) His solar power idea received serious consideration during the 1970s as a result of the 1973 oil crisis, but the technology of the time was not sufficiently advanced to make his concept a reality. Periodically, Glaser's idea is reexamined in the United States and elsewhere, and the further technology advances, the fewer technological and economic barriers remain in the way of its implementation.

There are now several competing proposals for space-based solar power. Each proposal is a different answer to a few fundamental questions. Should, for example, the satellites be placed in geosynchronous orbit or should the photovoltaic arrays be placed elsewhere—on the Moon, for example? Should the power be beamed to Earth at microwave frequencies or by laser beams? How big is the budget, and what will the cost of terrestrially based power be at the time the system is deployed?

flat ground so that a vertical stick casts no shadow—the amount of energy delivered per unit time per unit area of (flat) ground will be at its maximum. When the Sun is lower in the sky, the rays of the

The answers to these questions determine, in large measure, the characteristics of the design.

To convey a feeling for what is involved, a representative scheme involves placing satellites in geosynchronous orbit—that is, they are placed in an orbit that requires 24 hours to complete so that they are always directly above the same spot on Earth's surface—and they are outfitted with photovoltaic arrays that span about a square mile (about 3 km^2). The satellite would be stabilized so that the Sun's rays would always be perpendicular to the PV arrays. Even using today's photovoltaic technology, this configuration would generate very large amounts of power. The resulting energy would be broadcast toward Earth in a narrow beam using a wavelength within the 5.17–12.2 cm range, chosen because at these frequencies the atmosphere is reasonably transparent. The beam would arrive at the Earth's surface with an energy intensity of about one-sixth that of sunlight, an oft-quoted figure that makes the intensity of the energy sound more benign than it would actually be. (Workers at the site would need to limit their exposure.) The goal would be to provide gigawatts of power—but is it realistic?

Building a power station in space would require a new type of launch vehicle capable of making frequent trips into space at a much lower cost than today's technology allows. It would require significant advances in photovoltaic technology in order to make such huge arrays cost-effective to build and deploy, and it would require significant advances in robotics in order to build and maintain these huge, sophisticated structures. Evidently, it is a very expensive proposition. But the benefits of deploying these solar-powered orbiting generating stations may be worth the investment because they would provide direct access to an inexhaustible and continuous supply of power sufficient to meet the needs of the entire planet.

Sun deliver less energy per unit time per unit area, because the rays of the Sun are more spread out across the surface. The reason that the energy delivered to the site diminishes is primarily the result of

View of the Earth as seen by the *Apollo 17* crew traveling toward the Moon. Earth's geometry and rapid rotation greatly increase the difficulty of effectively harnessing solar energy on the surface. *(NASA)*

the change in orientation of the surface relative to the Sun's rays. The Sun, of course, continues to radiate energy at the same rate all of the time. Nevertheless, as the Sun's rays become more nearly parallel to the ground, the rate of energy delivered to a given patch of horizontal ground per unit time approaches zero.

The point on Earth at which the Sun is directly overhead at a particular time is called the subsolar point. At the subsolar point, the Sun's rays are exactly perpendicular to the ground. All other things being equal, the subsolar point receives more energy than any other point on Earth's surface because it is at this point that the Sun's rays are most concentrated. The further from the subsolar point one moves, the less power the Sun delivers per unit of ground. All other things being equal, a solar power plant located at a subsolar point will produce more power than one that is not.

As Earth rotates on its axis, the subsolar point continually shifts to the west. Over a 24-hour period, the set of all subsolar points forms a narrow band encircling the planet. Elsewhere on Earth, at those points where the Sun did not pass directly overhead, the maximum energy deliverable by the Sun to a flat patch of ground per unit time was less than at a subsolar point. In all cases, the

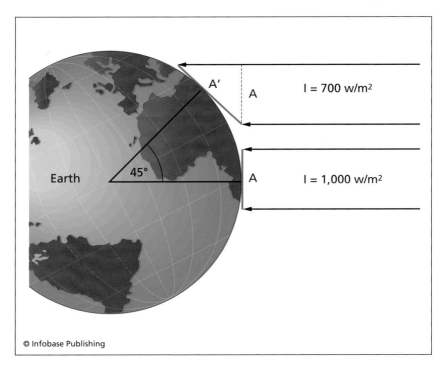

© Infobase Publishing

The more inclined the Sun's rays are to Earth's surface, the less energy is delivered per unit area per unit time. Forty-five degrees north of the subsolar point, the maximum amount of energy deliverable by the Sun per unit time per unit area has dropped by 30 percent.

maximum energy deliverable by the Sun to any piece of flat ground, regardless of location, varied over the course of the day. It was very small early in the day; it rose to a maximum when the Sun was at its highest point in the sky for that location, and finally, as night fell, the amount of energy delivered to the site decreased to zero. The intensity of sunlight varies, therefore, with both time and latitude.

There is still one more aspect of the geometry of the Earth-Sun system with important consequences for the utilization of solar power, namely the tilt of Earth's axis relative to the plane of its orbit. Earth's orbital path is flat in the sense that its orbit lies in a plane. Called the *plane of the ecliptic,* this plane also contains the center of the Sun.

Earth's axis of rotation is not perpendicular to the plane of the ecliptic. Instead, its axis of rotation is tilted about 23.5 degrees relative to a line perpendicular to the plane of the ecliptic. Loosely speaking, the axis of rotation pokes obliquely through the plane, and Earth maintains that orientation throughout the year. Over the course of a year, therefore, the intensity of the Sun's energy at Earth's surface varies with respect to latitude. In particular, the collection of subsolar point shifts according to the seasons. When it is summer in the Northern Hemisphere, the band of subsolar points for any particular day is located north of the equator. In fact, in the Northern Hemisphere on the longest days of the year, the band of subsolar points is located along latitude 23.5 degrees north, also known as the tropic of Cancer. When it is summer in the Southern Hemisphere, the band of subsolar points for any given day is located south of the equator, and on the longest day of the year in the Southern Hemisphere, that narrow strip of subsolar points is located along latitude 23.5 degrees south, also known as the tropic of Capricorn. To summarize: The length of the day and the latitude at which Sun's energy is most intense vary throughout the year because Earth's axis of rotation is not perpendicular to the plane of the ecliptic.

Taken together, all of the facts described in this chapter mean that the amount of solar energy available for conversion into electricity at any point on Earth's surface is highly variable. It varies

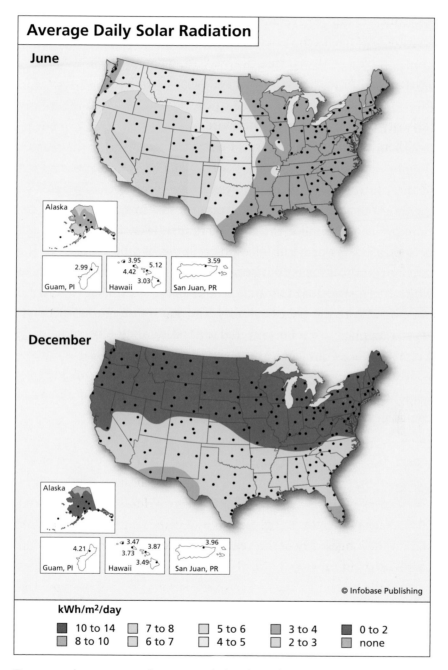

Average Daily Solar Radiation

June

Alaska

2.99
Guam, PI

3.95
4.42   5.12
3.03
Hawaii

3.59
San Juan, PR

December

Alaska

4.21
Guam, PI

3.47
3.73   3.87
3.49
Hawaii

3.96
San Juan, PR

© Infobase Publishing

kWh/m²/day

| | 10 to 14 | | 7 to 8 | | 5 to 6 | | 3 to 4 | | 0 to 2 |
| | 8 to 10 | | 6 to 7 | | 4 to 5 | | 2 to 3 | | none |

Two maps that represent the average daily solar radiation throughout the United States during the months of June and December. The dots on the maps represent the 239 sites of the National Solar Radiation Data Base *(NSRDB). (Source: NREL)* according to the season and according to the time of day; it varies according to the weather and according to air pollution levels. None of these factors is under the control of the plant operator. Whatever advantageous characteristics solar power stations may have—and there are several and they are significant—solar power is inherently intermittent. Unlike fossil fuel–fired power plants, which can be operated anytime, day or night, no matter the weather or the season, the power generated by a solar powered facility is in many ways out of the control of the operator, even when the plant is operated in the most efficient way possible. A natural gas–fired power plant, for example, can respond to demand. As demand rises, power levels can be increased. As demand falls, power levels can be decreased. The operator controls the output because the operator controls the rate at which fuel is supplied to the boiler. Solar plants are unable to respond to demand with the same flexibility. The maximum rate at which energy is supplied to a solar-powered generating station occurs during a certain time of day at a certain time of year provided the weather is good and the air is clear. These factors may or may not coincide with demand. Diminishing the effect of intermittence is one of the main challenges of solar power plant engineers and operators.

Photovoltaics

Photovoltaic (PV) technology converts sunlight directly into electricity. Improving the efficiency of photovoltaic technology and decreasing its costs are major research objectives at many laboratories, both public and private, around the world. The reason is simple: PV technology has the potential to convert more of the Sun's light into electricity than other competing technologies such as, for example, the more common *parabolic trough* systems, which are described in chapter 4. And PV technology is already common in a number of niche applications. Some calculators depend on PV cells for their power, for example, and PV cells provide power in situations where access to the grid is impossible. Other specialized applications exist. But despite decades of research, PV technology still makes a negligible contribution to the supply of electricity for the grid. Nor have many homeowners turned to arrays of PV panels as an alternative to the grid. To appreciate PV technology, it is

Mars Rover. Photovoltaics are ideally suited for use in the space program. *(NASA)*

important to understand both its successes and its failure thus far to compete with more conventional power generation technologies. This chapter begins by describing some of the relationships between light and PV-generated electricity. It goes on to describe specific PV technologies.

LIGHT VERSUS ELECTRICITY

Modeling light as a collection of waves has proven to be a very useful way of describing some of light's physical characteristics, but in order to better account for all of light's properties, scientists have found it necessary to also model light as a collection of particles. These particles are called photons. In the particle model, light can be described in terms of the energies of the individual photons of which it is comprised. The two descriptions complement one another. Light of a certain color can, for example, be described either in terms of its wavelength or in terms of the energy of its photons. Red light, which occupies that part of the visible spectrum with the longest wavelengths, consists of photons with energies of 1.7 electron volts—abbreviated 1.7 eV. (One electron volt equals 1.6×10^{-19}J.) And violet light, which occupies that part of the visible spectrum with the shortest wavelengths, consists of photons with energies of 3.1 eV. The photons responsible for violet light carry about 82 percent more energy than those responsible for red light. In general, the longer the wavelength of the electromagnetic wave, the lower the energy of the corresponding photons.

Modern PV cells are the basic units from which solar arrays are constructed—they are energy conversion devices. Made of special materials called *semiconductors,* they absorb photons with energy values within a certain range and convert the energy of those photons into electrical energy. They do not, however, convert all photons into electricity. Even in theory this is not possible. Instead, each type of PV cell absorbs some photons with energies within a range that depends on the characteristics of the PV cell in question. Crystalline silicon cells, which are one of the most common types of PV cells, absorb photons with energies in the visible light part of the spectrum as well as some photons lying within the infrared range. Other types of materials will absorb photons with energies within a different range of values, but the basic idea is the same for each type of PV cell. When a collection of photons impinges upon a PV cell, some photons are reflected, some are transmitted, and some are absorbed. Only the absorbed photons contribute to the production of electricity. A great deal of research is devoted to creating affordable photovoltaic cells that absorb as large a percentage of the available photons as possible. Typically, many common types of PV cells cannot, even in theory, convert more than half of all incident sunlight into electricity. Commercial PV cells generally convert much less than their theoretical maximum.

PV cells, also called solar cells, are the basic unit of a PV system. Most cells are small and produce at most a few watts of power even under ideal conditions. The PV cells are assembled into modules, collections of solar cells linked together to function as a single unit. Modules are the smallest practical units for a PV system. The modules are assembled into arrays, which are what the large PV panels are called. The size of an array depends on the amount of power required, the amount of sunlight available, and the amount of money the consumer is willing to pay. The more intense the light that strikes the cells, the more power flows from each module, and the smaller the array required to meet demand. If, however, the light is weak, the amount of power flowing from each module will be small, and as a consequence, the array must be large to generate sufficient power.

Because PV arrays are sensitive to the intensity of the light, they must be oriented to face the Sun. In theory, the best way to accomplish this goal is to mount an array on a mechanized frame that turns to follow the Sun across the sky. In practice, however, installing solar tracking devices greatly increases the costs of building and maintaining the system. As a consequence, arrays are usually mounted on stationary frames that in the Northern Hemisphere are turned south to maximize the amount of sunlight that shines on them. In the Southern Hemisphere, the arrays are oriented north to accomplish the same goal. The tilt of an array relative to the vertical depends on the latitude in which it is located. All of the geometrical considerations are calculated to make the best use of sunlight, the energy source from which the electricity is generated.

One of the distinguishing features of solar cells is the type of electricity that they produce. Electricity can flow in one of two ways. Direct current (DC) flows in one direction only, much the way water flows through a pipe. Alternating current (AC) regularly reverses directions. Batteries, for example, produce DC, and household current, the electricity that flows from outlets in North

American homes, is typically AC. Household current changes directions between 50 and 60 times per second. Solar cells produce DC current—that is, for as long as the Sun shines on the PV cells, direct current flows out from the solar array.

That solar cells produce DC as opposed to AC makes them unusual. Most electricity produced for the grid, whether from geothermal power plants, nuclear power plants, coal-fired power plants, wind turbines, the thermal plants described in chapter 4, or hydroelectric plants is produced as AC. Household appliances are generally built to run on AC, and so the electricity produced by solar cells must be converted to the same type of AC power available from the grid before it can be used. The device that converts DC to AC is called an inverter. Inverters are necessary; they also add to the cost of building and maintaining the unit.

Once a solar array has been installed together with the inverter and other accompanying hardware, the power that it produces is determined by local conditions. The weather, the amount of pollution in the air, and the length of the day, for example, all affect the intensity and duration of the local sunlight. The power that a solar array produces is not, however, determined by demand for electricity. There is no way of increasing the amount of power available from a given solar array if the power that it produces is insufficient to meet demand—that is, solar arrays produce power independently of demand. Often, they produce no power at all. At night, for example, when a PV array produces zero power, demand must be satisfied by using stored power or finding an alternative supply. Homeowners who depend on PV arrays, for example, may store power generated during the day in large batteries for nighttime use. When electricity is required during the night, electricity is drawn from the batteries, passed through an inverter, and used as needed. Because the Sun provides power intermittently, PV arrays can only produce power intermittently. Finding effective ways of dealing with the problem of intermittence is one of the major challenges for engineers involved in designing PV systems. The goal of PV research is to maximize the performance of PV systems within these limitations while minimizing the costs of their production.

TECHNICAL CONSIDERATIONS

To repeat: Solar cells are energy conversion devices, and as such they cannot produce more energy (in the form of electricity) than they receive (in the form of sunlight). Because PV systems are one of the most expensive ways of producing electricity, it is vital that they convert sunlight into electricity as efficiently as possible. This can be accomplished in several different ways, and the differing solutions in use indicate the challenges faced by engineers as they attempt to make solar arrays economically more competitive with conventional power sources.

One technique for making optimal use of PV cells involves increasing the intensity of the sunlight on the cells. This is accom-

Increasing numbers of home owners are turning to solar cells to meet part of their energy needs. *(Gray Watson and Rosemary McCrudden)*

plished by using lenses. A lens can focus light from a larger area onto a smaller area. By increasing the energy input for each cell, it follows that the energy output should increase as well. But the problem is more complex than that.

By adding a lens to each cell, the cost of each individual cell-lens combination is increased, but because (all other things being equal) the electrical output of each cell is increased, fewer cells are needed to produce the same amount of power. Since a cell always costs more than a lens, the decision to add a lens might appear to be a simple one. But all other things are never equal. Three factors make lenses problematic. First, the efficiency with which a solar cell operates depends on its temperature. The conversion efficiency drops as the temperature increases. Concentrating sunlight causes the cell temperature to increase and its efficiency to decrease. Part of the gain in efficiency attained by adding a lens can, therefore, be lost as the temperature of the cell rises. Second, a lens works best when the sunlight is perpendicular to its surface. In order to obtain the maximum benefit from lenses, the array must be mounted on a frame that automatically tracks the Sun across the sky. The tracking system increases both the installation and maintenance cost of the system. Finally, the value of the lens as a method of boosting power production depends, in part, on the quality of the light. Diffuse light can still be fairly bright, for example, but no lens will concentrate it. Even on a clear day, about 20 percent of the Sun's light is diffuse. In some areas, the percentage of diffuse light is much higher—even on a clear day. And on a cloudy day, a lens is useless. By contrast, an array of solar cells that does not use any concentrators (which is the technical name for lenses and lens-type devices) will usually generate some power even on cloudy days. Consequently, the decision of whether to use a concentrator depends on the economics of the particular installation and the weather at the site under consideration. There is no single optimal solution for every situation.

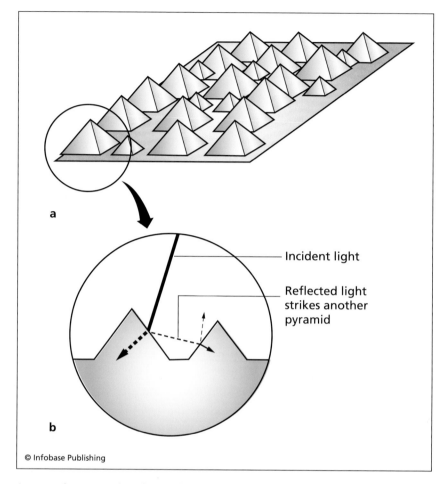

Incident light

Reflected light strikes another pyramid

a

b

© Infobase Publishing

A properly textured surface will absorb more light than a smooth surface.

Another way to increase efficiency is to decrease the amount of sunlight reflected off the surface of the PV cell. Reflected light generates no electricity, and smooth silicon, the most common material used to manufacture PV cells, reflects about one-third of all the light that strikes it. One solution is to add multiple antireflective coatings to the surface of the solar cell. These are chosen to decrease the reflection of light at certain key wavelengths. A second method of diminishing the tendency of the cell to reflect light is to texture

(continued on page 41)

(⏻) Photons and Electricity

Engineers have invented a number of different types of PV arrays, each with somewhat different characteristics. Practical PV arrays balance efficiency and cost. They must convert as much sunlight into electricity as possible at a cost that consumers will pay.

Solar cells have no moving parts. They can perform for decades and with little maintenance. As a general rule, the batteries, the inverter, and the rest of the system will wear out long before the solar cells stop functioning.

Crystalline silicon was the first material used to produce commercial products, an effort that dates to the 1950s. The process used to produce

(continues)

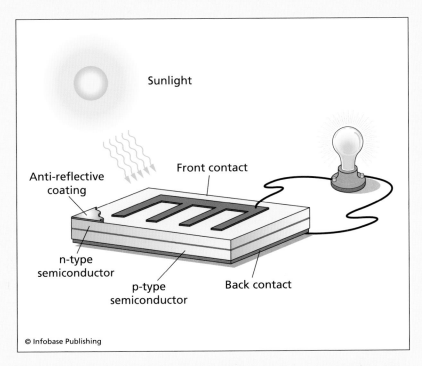

Sunlight

Anti-reflective coating

Front contact

n-type semiconductor

p-type semiconductor

Back contact

© Infobase Publishing

PV diagram. These devices convert sunlight into direct current electricity.

(continued)

these products has been refined, and the efficiency of the cells has improved, but crystalline silicon is still very expensive to manufacture. Amorphous silicon PV cells are a less expensive alternative. They are manufactured by depositing thin films, one on top of another, on an inexpensive substrate, or backing material, such as glass or plastic. Amorphous silicon technology and other so-called thin-film technologies are also usually less efficient than crystalline silicon. Other PV cells, built from even more expensive materials than crystalline silicon, are sometimes chosen because they result in higher-efficiency PV arrays.

All of these technologies use carefully crafted materials called semiconductors, made from so-called p-type and n-type materials. By placing the appropriate p-type and n-type material side-by-side an electric field is created that acts on free electrons—that is, electrons not bound to a particular atomic nucleus—and causes them to move in the direction of the electric field. This organized movement of electrons is electricity.

In the n-type layer, many electrons are only weakly held in place. When this layer absorbs a photon of sufficient energy, the energy carried by the photon may dislodge one or more electrons. Because these electrons are now loose in the electric field formed by the p- and n-type layers, they begin to flow in the direction dictated by the electric field. Each cell has electrical contacts that serve as the entrance and exit points for an electrical circuit through which the electrons flow. From the electrical contact on the front of the cell, they flow out of the n-type semiconductor, around the circuit where they might, for example, be harnessed to power a lightbulb or provide power for a computer or refrigerator, and finally, they return to the PV cell through the back contact, which is positioned on the p-type layer, to complete the circuit. (Unless the circuit is complete, the electrons cannot flow through it.) The energy that the electron acquired from the photon was dissipated by the work the electron did as it moved through the external circuit. Given enough time and sufficient light, it will eventually be reenergized by another photon and repeat the process.

(continued from page 38)
the surface so that it consists of numerous very fine prism-shaped bumps. Light that strikes one angular surface is either absorbed or reflected toward another angular surface. A carefully textured surface will cause a ray of light to strike the surface multiple times. Each time a ray of light encounters an angular surface, most of the remaining light is absorbed, and the remainder is directed against still another tiny piece of surface material. In this way most of the photons are absorbed, if not at one location on the surface then at another.

Because the properties of sunlight depend very much on the location under consideration, it is important to have detailed information about the "insolation," the total amount of energy delivered by the Sun per unit area per unit time. Insolation values depend on how they are calculated—per month or per day, whether the surface on which the light falls is horizontal or turned toward the Sun, and so on. The U.S. government has compiled a variety of databases containing detailed information on insolation values. It is expressed in the form of maps, charts, and graphs, and has been obtained from terrestrial measurements and satellite measurements. The quantity of information is astonishing. It is also vital when evaluating the potential of solar cells for any given location.

NET METERING

Money and electricity can be converted one into the other. Money is converted into electricity when consumers pay their utility bills, and electricity is converted into money when power producers sell the electricity that they produce. This is the idea behind *net metering,* a method for optimizing the use of residential solar arrays.

It is often the case that a residential PV system produces power when it is least needed. During the day, when the Sun is shining and the PV system is producing its power, those who live in a home are usually elsewhere. They may be at work or at school, for example.

They are often not using the electricity generated by their very expensive PV array. Absent its occupants, home appliances generally use little electricity during times of peak solar production.

The most obvious way to make use of the electricity produced by the PV array is to store it in batteries. But battery storage systems are expensive to install; they require maintenance, and add one more level of complexity to a PV system. If a PV system is to be attractive to potential buyers, it must be a reasonable alternative to the grid, which, in contrast to PV-generated electricity, requires little more from the consumer than to plug in each appliance and flip a switch. One alternative to battery storage, called net metering, makes use of the observation that electricity and money are routinely converted one into the other. Net metering works by setting up the PV system so that when the array produces more electricity than is required, the surplus is injected into the grid. In this way, a PV-powered home may, at times, contribute to the electricity supply rather than the electricity demand. The amount of power that flows into the grid from a residential system is not large, and so utilities have found it convenient to use meters that run forward when electricity is drawn from the grid and backward when electricity is injected into the grid. If more electricity is injected into the grid over a billing period than is drawn from the grid, the home owner receives a credit. Otherwise, the home owner receives a bill. The two-way meter allows the owner of the PV system to convert electricity into money and vice versa. Under net metering the grid can be visualized as a sort of storage device, where electricity is stored in the form of a billing credit.

Not every utility allows net metering, and historically many utilities that currently permit net metering were forced to accept it. From the utilities' point of view, there is little economic justification for allowing net metering. Owners of residential PV systems benefit from the electricity infrastructure provided by their respective utilities. Indeed, they could not sell their surplus power without it, but they contribute nothing to the upkeep of the system even as the utilities assume the costs of maintaining the home owners' grid connection. Further, by permitting power to flow into the grid in a way that is only as predictable as the weather, the utilities' task of maintaining the system voltage within prespecified limits is made more complicated. In effect, the owners of residential PV systems receive a substantial benefit, the partial or total elimination of their monthly utility bill, and the utilities incur additional costs even as revenue goes down. From the utilities' viewpoint, net metering can be a money loser.

Net metering has failed to make much of an impact on the energy sector so far because the cost of installing a PV system is high enough to discourage most home owners from considering it. If costs come down significantly—or if electricity prices go up significantly—net metering may become more common, bringing with it important changes to the way that electricity is produced and consumed.

Heat Engines and
Solar Power

Concentrating solar power (CSP) systems are the solar-powered alternative to photovoltaic (PV) systems. CSP systems operate by first converting solar energy into heat and then converting the heat into electricity. Neither type of power generation technology has been widely adopted. As previously mentioned, all types of solar power produce less than one-tenth of 1 percent of total electricity production in the United States, with similar percentages in other large economies. But between the two technologies, PV and CSP systems, much more electricity is produced with CSP technology, and the reason is cost. Although CSP produced electricity is more expensive than electricity produced using more conventional technologies, it is a bargain compared with photovoltaic technology.

This chapter describes the basic principles that govern the design of CSP systems. It also describes a few of the more successful designs, as well as some advantages and disadvantages of CSP systems.

Solar Two on a sunny day *(TREC-UK)*

CONVERTING THERMAL ENERGY INTO ELECTRICAL ENERGY

CSP systems are heat engines, machines designed to convert thermal energy into work. Because they are heat engines, CSP systems are, in many ways, unremarkable. They operate according to the same basic principles as other heat engines, principles that would have been familiar to engineers working 100 years ago. CSP systems share many of the same components as coal-fired, natural gas–fired, nuclear, and geothermal power plants. The components of CSP systems are probably most closely related to those of geothermal power plants that operate on a binary cycle. (Binary geothermal power plants and the principles that govern their operation are described in chapter 10.)

As with more conventional electricity generating stations, the purpose of CSP systems is to spin a generator, a device that converts the motion of a turning shaft into electricity. This is accomplished in the following way:

1. The Sun's energy is concentrated with the help of mirrors. This energy is used to heat what is called the primary fluid, usually a type of oil or a molten salt.

2. The heated primary fluid is pumped to a radiator-like device called a heat exchanger, a device that permits heat to flow from the hot primary fluid to a cooler secondary fluid. The heat exchanger permits only the exchange of heat. The fluids do not mix. For CSP systems the secondary fluid is usually water.

3. As heat flows from the primary fluid to the liquid water, the water turns to high-pressure steam.

4. The high-pressure steam is pumped from the heat exchanger to a release valve. As it passes through the release valve, it enters into a region of lower pressure and expands rapidly. The valve directs the expanding steam against the blades of a turbine, a device designed to convert the straight-line motion of the expanding steam into rotary motion. The turbine spins in response to the force of the expanding steam. (Meanwhile the cooled primary fluid, from which thermal energy was removed at the heat exchanger, is pumped back to the solar collectors to be reheated.)

5. As the turbine spins, it turns the shaft that drives the generator, which produces electricity.

6. Once past the turbine, the steam is directed to a condenser, where it cools and reverts back to the liquid phase.

7. The liquid water is pumped from the condenser back to the heat exchanger, where the cycle begins again.

In common CSP systems, the primary fluid (oil or molten salt) must absorb a great deal of thermal energy without changing into a vapor. Typically, these fluids are heated until their temperature lies between 700°F and 800°F (370°C–430°C).

A second difference between CSP systems and most other heat engines is that CSP systems depend on a very diffuse source of energy to provide them with power. By the time the Sun's rays reach Earth, they are spread thinly, and by the time they pass through the atmosphere, they are weaker still. As a consequence, while the Sun provides enormous amounts of energy to Earth, when averaged over a 24-hour period the Sun does not provide very much energy at any particular location. In order to compensate for the comparative weakness of the Sun's rays, solar plants cover large areas of ground with mirrors that concentrate the sunlight and raise the temperature of the primary fluid until it is high enough to power the plant. By way of example, Nevada Solar One, located in Boulder City, Nevada, and one of the world's newer and larger solar power plants, uses 350 acres (160 hectares) covered with 182,400 mirrors to generate a modest 64 *megawatts* of electrical power. Each mirror has a surface area of 20 square feet (1.9 m²). (By contrast, a single large nuclear reactor will generate approximately 1,000 megawatts. When both facilities are operated at full power, therefore, the nuclear plant generates almost 16 times as much power as the Nevada facility. Also, a modern nuclear plant will produce power about 90 percent of time and in all sorts of weather, while the Nevada facility will generate power much less than half of the time and only during good weather.)

Three different CSP technologies are currently in use. The first type of CSP design, which is also the technology used at the Nevada Solar One facility, uses parabolic troughs to collect energy. A parabolic trough is designed so that when it faces the Sun directly, all of the rays of light that strike the curved surface of the trough are reflected so that they pass through a line that runs along the trough's central axis. This is the "concentrating" part of concentrating solar power technology. A pipe is installed so that it coincides with this line. The pipe is filled with the primary fluid. As the fluid courses through the pipe, it absorbs increasing amounts of thermal energy.

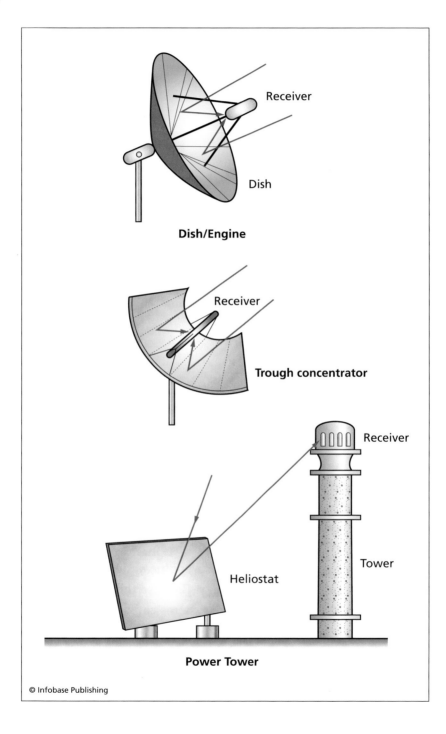

Receiver

Dish

Dish/Engine

Receiver

Trough concentrator

Receiver

Tower

Heliostat

Power Tower

© Infobase Publishing

(opposite page) Although the dish/engine design is the most efficient at converting solar energy into useful working energy, the trough concentrator is the most widely used, and the power tower was the first design to incorporate thermal storage.

most efficient of the three CSP technologies. In practice, however, dish/engine systems have been both expensive to build and expensive to maintain, and they have yet to be widely adopted even by the modest standards of the solar power industry.

ADVANTAGES AND DISADVANTAGES OF CSP TECHNOLOGY

CSP technology is only advantageous or disadvantageous relative to other power generation technologies. While some solar power supporters champion these technologies for philosophical reasons, the principal interest of most investors is the profitable production of electricity. The particular technology used to produce electricity—whether it is geothermal, solar, coal, natural gas, hydropower, or something else—is just a means to an end. In general, solar energy must compete with other power production technologies. Consequently, it is important for solar energy advocates as well as those who advocate for other technologies to impartially determine the advantages and disadvantages of each technology.

With respect to the issue of cost, it has already been noted that CSP technologies are significantly less expensive than photovoltaic power generation technologies. This fact, alone, is enough to explain why more CSP systems are generating commercially significant amounts of power than are PV systems. But despite their economic advantage relative to photovoltaic systems, CSP systems are more expensive and less reliable than many conventional technologies. In particular, they are still too expensive to deploy without significant government subsidies.

(continued on page 54)

Storing Solar Energy

One of the main objections to solar power systems is that their energy source, sunshine, is intermittent. This can be a severe constraint on their usefulness, because demand for electricity does not always coincide with sunny days. A plant that produces power only intermittently and without regard to demand has only modest economic value, and currently, there is no good way of storing large amounts of electricity directly. There are, for example, no giant batteries. Engineers have, therefore, searched for other ways of storing energy, and one way that shows particular promise is to convert solar energy into thermal energy and to store the thermal energy for later use.

Currently, the most promising technology for storing thermal energy involves a simple-sounding modification of existing designs: Instead of using all of the thermal energy as it is generated, some of the hot primary fluid is saved for later use. The key to a successful design is to choose a fluid capable of holding a great deal of thermal energy. Once heated, the primary fluid is stored in a specially insulated tank until it is needed. This design was first implemented at the Solar Two prototype facility in Barstow, California.

Molten salt can be heated to high temperatures. The Solar Two facility heated its molten salt primary fluid to about 1,000°F (540°C). A practical design must heat the primary fluid faster than it is needed to maintain power output and then store the surplus. The thermal energy stored in the salt can be used to produce electricity hours or even, in theory, days later.

In designing such a facility, plant engineers must make hard decisions about how to use the plant. There is a balance to be struck between the power output of the plant and its storage capacity. If the plant owner wants a plant with a large power output, it will have to accept a small storage capacity, because in order to produce large amounts of power most of the thermal energy will have to be used as it is generated. If, conversely, a plant owner wants a power plant that can produce energy over an entire 24-hour cycle, it will have to accept a much more modest power output from the plant, but that power output can be maintained over a much

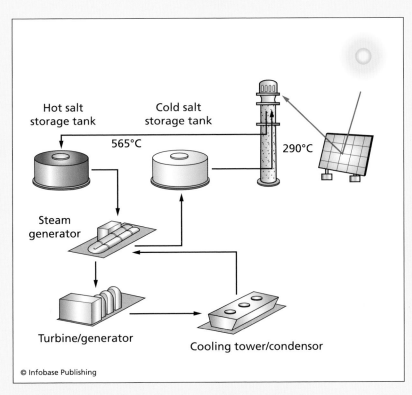

Diagram of a molten-salt CSP plant with thermal storage

longer period of time. For example, a power tower could be designed with little or no storage, but with the ability to produce 200 megawatts (MW) of electricity for four hours per day. By contrast, a plant with the same number of heliostats but outfitted with a large storage tank could be designed to produce a maximum 33 MW of electricity at full output, but by drawing on the energy in storage, it could maintain its 33-MW output throughout an entire 24-hour cycle. Both the 200-MW power plant and

(continues)

(continued)

the 33-MW power plants produce approximately 800 MWh (megawatt-hours) of electricity per day—that is, 200 MW × 4 hours is approximately equal to 33 MW × 24 hours. It is up to the power producer to determine which it values most: a higher maximum power output available for fewer hours per day or a lower power output available for a longer period of time. It cannot choose both. Power plants are energy conversion devices. If they both collect the same amount of solar energy and convert the same percentage of that energy into electricity, they will produce the same amount of electricity for each thermal "charge."

The first commercial power tower plant with thermal storage has recently been completed in Spain. It has an output of 17 MW of electricity. Called Solar Tres, it has facilities to store sufficient thermal energy to main output for 15 hours. Parabolic trough plants using storage tanks are also in various stages of development.

(continued from page 51)

An important advantage of CSP systems is that they can be deployed now. At a time when newspapers and magazines enthusiastically report on the future of technologies that have yet to be developed, CSP systems provide a proven way to produce renewable power. A power producer that wants to acquire experience producing renewable power can buy a CSP system from one of several vendors. Many utilities have an obligation to buy a specified fraction of their power from renewable sources, and CSP systems can be used to satisfy that requirement because their operation produces no greenhouse gases. They are as pollution-free as any source of power currently available.

CSP systems, however, do not make very efficient use of sunlight. Part of the problem arises from the fact that unlike photovoltaic power, which converts sunlight directly into electricity, CSP technology converts solar energy to thermal energy prior to the production of electricity. This extra conversion results in significant losses. Consider, for example, a parabolic trough plant. While each design has its own performance characteristics, an average plant might convert between 30 and 60 percent of the sunlight that strikes the mirrors into thermal energy. The steam produced is only of moderate temperature and pressure (relative to more conventional power plants), and as a consequence, only about 20 percent of the thermal energy is converted into electrical energy. Therefore, of the 100 percent of solar energy that strikes the mirrors of the plant, between 6 and 12 percent ($0.30 \times 0.20 = 0.06$ and $0.60 \times 0.20 = 0.12$) is converted into electricity. Low conversion efficiencies explain why large-looking CSP facilities have relatively small outputs. By way of illustration, although Nevada Solar One occupies about 350 acres, it has a maximum output of 64 MWe. Doubling the plant's output would mean doubling (at least) the amount of land covered by the reflectors. Large commercial solar facilities will require proportionately large land resources.

CSP plants are currently the best solar technology for producing commercially significant amounts of electric power, but as illustrated by the amount of land they require, these plants can have significant effects on the environment. There is no easy way to generate large amounts of electricity.

Two Other Important Solar Technologies

There is more to the solar energy industry than the production of electricity, but electricity production is extremely important because of all forms of energy, electricity is the most versatile: It can be used to power appliances and electronics, to provide the energy for cooking, lighting, and communication, and to power heating and cooling systems, simple or complex. It is also the easiest form of energy to transmit over long distances. Converting sunlight into electricity in a cost-effective manner has, however, proven far more difficult than was predicted by many 1970s-era writers, who confidently predicted a major role for solar energy by the year 2000.

There are a number of important nonelectric applications for solar energy. The solar water heater, for example, is a relatively simple technology with important environmental implications. A modest number of systems have already been deployed. This chapter describes the basics of this technology and why it is important.

Sun's rays through forest canopy in Yosemite. Nonelectric applications of solar energy are also important. *(Krishnaram Kenthapadi)*

Passive solar design also uses the energy in sunlight without first converting it into electricity. So-called *green architecture,* building designs that make more efficient use of local sources of energy, depends in an essential way on passive solar design. The impact of passive solar design on energy consumption patterns over the medium term will be relatively small since most buildings that will be in use over the next few decades have already been built. Nevertheless, the effects of better architectural designs are cumulative, and the effects of green architecture over the long term may prove to be very important. This is the second topic of this chapter.

SOLAR WATER HEATERS

Solar-powered water heaters can significantly reduce the environmental consequences of the energy-intensive lifestyles characteristic of individuals in developed nations. To understand why, consider how heat affects water.

When thermal energy is transferred to a liquid—and provided the liquid is not changing phase from, for example, a liquid to a gas—the temperature of the liquid will increase. This simple fact conforms to everyone's daily experience. (During a phase change, the temperature of the liquid does not increase as heat is transferred. Instead, the rate at which thermal energy is transferred to the liquid affects the rate at which the phase change occurs.) Less appreciated, however, is that when two liquids of equal mass but differing chemical composition absorb the same amount of thermal energy, one liquid will usually show a larger change in temperature than the other. Some liquids require less heat to warm up. This is most easily illustrated by an example: Suppose that a unit mass of water and a unit mass of ethylene glycol, a chemical commonly used in automotive antifreeze, both absorb one unit of thermal energy. The change in temperature of the ethylene glycol will be almost twice as large as the temperature change of the water. To be clear: This is true even though the samples have identical masses and absorb identical amounts of thermal energy.

Specific heat, which is usually represented with the symbol c_p, is the concept used to describe the way a sample of material—liquid, solid, or gas—responds when thermal energy is transferred to or from the body. At atmospheric pressure and at 60°F (15°C), the temperature of one pound of water will increase about one degree Fahrenheit with the addition of one *Btu* of thermal energy. This is summarized with the equation c_p (water) = 1Btu / lb · F. Using metric units the preceding statements can be summarized as c_p (water) = 4,184J / kg · C. (The exact value of the specific heat of water varies slightly with temperature, but between 32°F [0°C] and 212F° [100°C], the specific heat of water at one atmosphere pressure is almost constant. For most practical applications, therefore, there is little harm in assuming that the specific heat of liquid water is constant.) By contrast, the specific heat of ethylene glycol is 0.53 that of water so that

$$c_p \text{ (ethelene glycol)} = 0.53c_p \text{ (water)} = 0.53Btu\ /\ lb \cdot F = 2,218J\ /\ kg \cdot C$$

In fact, compared to most liquids, liquid water has a very high specific heat. It can absorb and emit large amounts of thermal energy without undergoing large temperature changes. This is why, for example, air temperatures near large bodies of water tend to be more moderate than air temperatures further inland. The water moderates the weather by absorbing heat as air temperatures begin to rise and emitting heat as air temperatures begin to cool, although the temperature of the water itself changes only slightly over the course of the day.

The specific heat of a sample of a liquid is related to the change in temperature of that liquid by the equation

$$Q = c_p m (T_f - T_i) \qquad\qquad \text{(Equation 5.1)}$$

The letter Q stands for the thermal energy transferred in or out of the liquid. The symbols c_p and m represent the specific heat of the liquid and its mass, respectively, and the symbols T_f and T_i are the final and initial temperatures of the liquid, respectively. If T_f is less than T_i, then Q is negative, which indicates that thermal energy was removed from the body rather than transferred into it. (In this chapter, liquids are emphasized, but the same ideas also apply to gases and solids.)

Because water has a high specific heat, heating water is a very energy intensive activity, one of the most energy-intensive activities that most people undertake during the course of an average day. By way of example, heating one gallon (3.785 liters) of water, initially at a temperature of 60°F (16°C), until it begins to boil requires about 1,300 Btu (1.4 MJ). (This can be verified using equation [5.1] and the fact that one gallon [3.785 l] of water weighs about 3.785 kg. Use metric units to perform the computation.) Another way to think about this result is that boiling water under the conditions just indicated requires the same amount of energy as is required to operate a 60-watt bulb for more than six hours, assuming that all of the heat used to raise the water temperature is transferred to the water. In practice, of course, all of the heat is never transferred to the water, and heating a gallon of water on a kitchen stove releases enough heat into the kitchen to heat the kitchen as well. The figure 1,300 Btu (1.4 MJ) is, therefore, a lower bound, and heating a gallon of water actually requires much more heat than the preceding computations indicate. But even the 1,300 Btu figure is a substantial amount of energy.

What the preceding discussion demonstrates is that the direct use of solar energy to heat water (instead of natural gas, electricity, or oil) can make an important change in energy consumption patterns, and because solar-powered water heaters generate little or no pollution, they can also reduce the environmental impact of one of the commonest and most energy-intensive of everyday human activities.

Broadly speaking, solar hot-water systems consist of two components. The first component is a solar collector, which consists of one or several flat plates designed to absorb the Sun's rays as efficiently as possible. Embedded within these are numerous small pipes through which fluid flows. Solar energy is absorbed by the collector and is converted into heat, which is absorbed by the fluid as it passes through the tubes, causing the fluid to become warm. If the fluid is water, then the water is piped to a carefully insulated storage tank, where it is held until it is needed. Often, however, the "working fluid," the term for the fluid that circulates through the solar collector, is not water. Instead, a harmless mixture of glycol and water is often used. The glycol prevents the mixture from freezing. The working fluid flows in a closed loop—closed in the sense that it remains permanently within the loop and, in particular, does not mix with the household water supply. Once the working fluid has been heated, it flows through a heat exchanger, a radiator-like

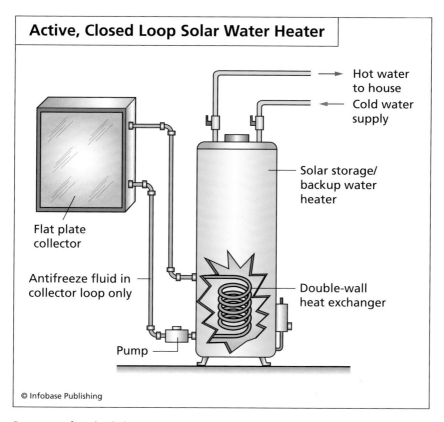

Active, Closed Loop Solar Water Heater

Hot water to house

Cold water supply

Solar storage/ backup water heater

Flat plate collector

Antifreeze fluid in collector loop only

Double-wall heat exchanger

Pump

© Infobase Publishing

Because of its high heat capacity, heating water is a very energy-intensive activity.

device that permits the flow of heat out of the working fluid and into the home's water supply. At the heat exchanger, the temperature of the household water increases and the temperature of the working fluid decreases. The working fluid is then pumped back to the collector to be reheated.

Some systems depend on pumps to move the working fluid, and some depend on the fact that a sample of warm fluid is naturally more buoyant than the same fluid that has been cooled. Engineers make use of the natural buoyancy of the warmer fluid to create a system in which the fluid circulates without pumps. Circulation

⏻ Demand Management

Demand for electricity is not constant. It is, on average, higher during the day than it is late at night and higher Monday through Friday than it is during weekends and holidays. The minimum level of demand is called the *base load power* requirement. Coal plants, nuclear plants, and (where they are available) geothermal power plants are used to supply base load power. These technologies produce electricity reliably, inexpensively, and continually for extended periods of time. They are, however, ill-suited for rapid start-ups or quick shutdowns. As demand rises above the base load power requirements, other types of power plants are brought online. They provide what is called peak power, and they may only be used for a few hours at a time. Natural gas–fired power plants, for example, are commonly used for this purpose, because these plants are reliable and can be started and stopped relatively quickly. Solar power plants can also contribute to *peak load* power production since they tend to be most productive when demand is highest. Peak load power is generally sold at a premium since peak load power plant owners must earn their profits from the few hours per day that they operate their plants. Peak load power production is inefficient in the sense that meeting peak demand requires the construction and maintenance of plants that remain idle most of the time.

Because peak load power tends to be more expensive, one way of reducing the price of electricity would be to reduce fluctuations in demand. Reducing demand fluctuations is different from conservation. Conservation emphasizes reductions in consumption. So-called *demand management* emphasizes shifting consumption from periods of peak load power production to off-peak hours. Graphically, demand management seeks to flatten power production curves, the curves that show how power production varies with time. At present, power production curves rise and fall fairly dramatically. By flattening the power production curves, the same amount of energy may be

of this type is called natural convection, and not surprisingly, such systems are cheaper to install and operate. But flow rates in systems that depend on natural convection tend to be lower than in those systems that produced over the course of a day or a week, but it would be produced at different times—more at night, say, and less during the day.

In order for the strategy of demand management to succeed, it must modify consumer behavior. Utility customers must be convinced to change the times that they consume electricity so that their demands can be met with less expensive base load power. Accomplishing this goal requires the installation of more sophisticated electric meters, devices that measure not just the energy consumed but also the time at which it was consumed. This type of meter enables the utility to charge customers on a sliding scale: higher fees for electricity used during peak periods and lower fees for off-peak usage. In theory, if electricity were priced according to the time that it is consumed, more electricity would be consumed during off-peak periods and less electricity would be consumed during peak periods. This would "flatten out" the demand for electricity, and utilities could buy more electricity under long term contracts from less expensive base load power producers. Instead of having more plants to generate peak load power, increased base load demand could be met with a modest increase in the number of base load power plants. Consumers would be compensated for the inconveniences involved in changing their consumption patterns with lower utility bills.

With respect to solar power, demand management might reduce interest in large-scale solar powered generating stations because they are best suited for peak power production. Residential solar applications, by contrast, may benefit from demand management because they produce electricity and heat at exactly the times when electricity and heat would be most expensive. Consequently, the time required to recover the cost of residential solar power devices would be shortened, and they would become more attractive as investments.

use pumps. The design of the system should reflect the needs of the owner. There are now a wide variety of solar hot-water system designs, some of which are more efficient than others. The buyer is compensated for the costs incurred in purchasing, installing, and maintaining the system by the corresponding reduction in energy costs and (usually) generous government subsidies.

As with all solar technologies, the value of any solar water heater depends on the latitude at which it is deployed as well as the weather and a number of other local conditions. But even in locations where a solar water heater fails to raise the temperature of the water sufficiently for residential use without additional heating, it may still be worthwhile to deploy such a system as a preheater. Economically and environmentally, the cost of heating water is high.

GREENER ARCHITECTURE

Until recently, most modern office buildings were comprised of a light steel frame clad in acres of glass. These buildings trap solar energy almost as if they were sealed cars on a hot summer day. In these buildings, when the Sun rises, the air conditioning turns on, and when the Sun sets, the air conditioning turns off, and the heating system turns on. The flow of energy in and out of the building through the sealed windows can only be counteracted with the most energy-intensive technologies. Similarly, most homes are still built so that the side with the front door and the biggest window is parallel with the road. Both types of buildings are, more often than not, still sited without regard to the position of the Sun. These standard commercial and residential designs reflect a general lack of concern about the environmental and the economic costs of power production. They were built during an era of cheap power. As energy costs continue to climb and the environmental costs of old consumption patterns become increasingly apparent, building designs have begun to change as well. These new energy-conscious architectural designs are often described as "green."

Some aspects of green architecture are easy to identify. Green buildings make use of high-efficiency insulating materials; they may use solar panels to heat water or photovoltaic arrays to reduce their utility bills. But these technologies do not by themselves make

Passive solar building, Tinicum, Pennsylvania. A building designed to make maximum use of local conditions can achieve considerable savings in energy costs. *(Cusano Environmental Education Center, John Heinz National Wildlife Refuge)*

the design of a building green. They can, in fact, be retrofitted on almost any structure. A green design is different in more fundamental ways.

One of the most important factors in the energy budget of a building is the cost of heating and cooling. One way to begin to reduce these costs is to site the building with respect to the Sun rather than the road. If heating is the primary concern, the windows can be sized and placed so that they allow as much of the Sun's energy to enter as possible. In areas where cooling is the primary concern, windows can be placed so as to minimize the amount of solar heating. Through the use of overhangs and other devices, it is possible to optimize the amount of sunlight entering the structure during the winter when the Sun is low in the sky and to minimize the amount of sunlight entering the building during the summer when the Sun is high in the sky.

The interior surfaces—walls and floors—can be made of special materials that can serve as heat sources and sinks. When sunlight shines on a floor, for example, the floor can be constructed so that it absorbs the thermal energy. The floor can then heat the room after sunset.

There are a number of other design principles that are important in green architecture. Some of these principles were incorporated into ancient buildings, so the ideas are not new, but today they represent a significant break with more recent practices. The use of natural lighting, for example, is a frequent goal of green building design. At one time, effective natural lighting meant better lighting; now, when natural lighting is no longer better than artificial lighting, effective natural lighting means a reduction in energy usage. But making use of natural lighting means siting the building correctly with respect to the Sun. It means using energy efficient windows, and it also means creating a design in which light can diffuse far enough into the building to make a difference. These ideas, while not universally embraced, are gaining popularity.

Ventilation costs, both economic and environmental, may be reduced with green designs. Naturally ventilated spaces that rely on convection driven by the Sun have been incorporated into buildings, large and small. Designing a system that provides fresh air throughout all, or at least part, of a building requires that the building be designed for the location in which it is erected. The Hawaii Gateway Energy Center in Kailua-Kona, Hawaii, for example, uses the Sun to heat a copper roof. The roof heats air within the building, causing it to rise through vents. The rising air draws cooler fresh air from the building's exterior through ducts located beneath the floor. The system requires no electricity to accomplish this process. The building incorporates a number of other innovations that use solar energy and cool water from the nearby ocean. The Gateway Energy Center is an award-winning example of green architecture, but it is not a design that will work away from the site on which it was built.

One more example: At the University of Utrecht in the Netherlands, one building, called the Minnaert Building, is designed to capture rainwater and funnel it to a nearby pond, and during the summer the water is circulated through pipes in the roof to cool the building.

Green designs are often local in nature—that is, each building must be fitted to its surroundings: an ocean, a pond, a given amount of sunlight, a given average cloud cover, and so on. In particular, a design that works well in one location may not fare as well in another. The intensity of the sunlight or the length of the day may be different; a building located in a valley may have different performance characteristics from one located on a nearby hill; neighboring buildings or even trees may interfere with the available sunlight, or the weather may differ between two otherwise similar sites. Individualized designs raise costs, and increased costs mean increased prices.

In the European Union and especially in the Netherlands and in Germany, where green architectural firms have found their biggest market, cost is, not surprisingly, an important issue, but it is not always the deciding issue on which design decisions are made. "In Europe, the guidelines tend to have to do with broader organizational ideas. Energy consumption, the organization of the workplace, urbanization—they're all seen as interlinked," said Thom Mayne of the architectural firm Morphosis in a May 2007 interview for the *New York Times*. Organization and urbanization may be important, but there is not an unambiguous way of assigning costs to these characteristics. In North America, where mandatory energy efficiency standards are lax or absent, customers have been slower to adopt green architectural designs unless, of course, the economic reasons for doing so are compelling.

Green architectural concepts will not change the landscape anytime soon. Most buildings have long lives, which is to say that most buildings that will be in use 10 years from now have already been built. The same is true of homes in developed neighborhoods. Green commercial buildings and green residential buildings will, therefore, make up only a small percentage of the total number of government, commercial, and residential building for many years into the future. But if the incorporation of green designs into the architectural landscape is slow, it is also cumulative. Eventually, provided these ideas remain important to designers, the effect of these designs on energy consumption may be substantial, but if the goal is to effect rapid change in consumption patterns, one must look elsewhere.

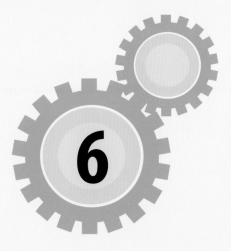

Economic and Environmental Consequences of Solar Power

All electricity produced for use on the grid is the same whether produced by a solar, geothermal, coal-fired, nuclear, natural gas–fired, or hydroelectric power plant. Utility customers cannot determine the origin of the electricity from the way that the devices that consume it operate. Over the short-term, utility customers are sensitive only to price. The price of the electricity is transparent; its origin is not. Consequently, technologies used to generate power are, over the short-term, indistinguishable if they all sell power at the same price. But price is very much dependent on the technology used to generate the electricity. One goal of this chapter is to examine some of the factors affecting the price of solar power.

Longer-term experience has shown that utility customers can also be affected by the way that electrical power is produced. In

Singapore at night. Worldwide demand for electricity is already very large, and it is growing rapidly. *(Northwestern University, Kellogg School of Business)*

certain parts of the United States, for example, the consumption of freshwater fish is a health hazard because their bodies contain elevated levels of mercury, a toxin that was emitted by coal-burning power plants located hundreds of miles upwind. And every fossil fuel–fired power plant releases carbon dioxide into the atmosphere with long-term effects on the planet's average temperature. Many other examples exist. This chapter also describes some of the environmental effects of solar power.

MATCHING SOLAR SUPPLY WITH CONSUMER DEMAND

The supply of solar energy is intermittent. Intermittence arises from one of two sources: Earth's rotation and the weather. One source, Earth's rotation, is completely predictable. It is, therefore, known with certainty that, as a result of Earth's rotation, a solar power plant will receive no energy from the Sun for a significant portion of each

24-hour period. The second source of intermittence, the weather, is less predictable. Because the weather cannot be accurately predicted far in advance, neither can the daytime output of a solar-powered plant. (Contrast solar output with the output of geothermal power plants, which typically operate at full power about 95 percent of the time.) Despite these drawbacks, solar power plants can make an important contribution to the energy supply. To understand why, it is necessary to understand how power production is shared among differing technologies.

Utilities generally enter into long-term contracts with so-called base load power producers. Base load power is the minimum amount of power that a particular system requires. Base load power requirements are highly predictable and consist of the power used by, for example, essential services and certain 24-hour per day manufacturing concerns. This power is manufactured by generating stations that can operate reliably and continuously for long periods of time. Geothermal power producers typically supply base load power as do nuclear plants, coal plants, some hydroelectric, and some natural gas–fired power plants. Solar energy would usually be a poor choice for base load power because sunlight is an intermittent energy source.

But there are often fluctuations above the base load power requirements. Spikes in demand, the dates of which are sometimes difficult to predict far in advance, may occur in response to factors such as weather. When faced with hot weather, for example, many homes and businesses turn on air conditioning units, and air conditioning is a very energy-intensive technology. Everyone knows that the hottest days generally occur during the summer, but no one knows very far in advance which summer days will be particularly hot. Other fluctuations in demand are more easily predicted: Demand is higher during the day Monday through Friday than it is during the day Saturday and Sunday, and demand is higher at three o'clock in the afternoon than it is at three o'clock at night.

Fluctuations above the base load requirements are called peak load power. (Sometimes fluctuations above the base load requirements are more finely classified into intermediate and peak, but classifying all demand above base load as peak load is sufficiently accurate for this discussion.) Peak load power is manufactured with power plants that are easier to rapidly start and stop. Natural gas–fired plants are often the preferred way to meet peak demand, because these plants are reliable and can be turned on and off relatively easily. (By contrast, nuclear and coal-fired plants are poor choices for rapid start-and-stop usage.) As already mentioned, natural gas can also be used to meet base load requirements, but because of the high cost of natural gas, they are increasingly used for short-term peak load power production.

Solar-powered plants can also be used to meet peak load power requirements. (Most solar power stations are thermal, but the same general remarks apply to PV stations, although few commercial-scale units exist at present.) Solar power stations are reasonably well suited to this sector of the power-generation market. They tend to produce power during the heat of the day, when demand tends to peak, and they, too, are relatively easy to start and stop. Night is, therefore, no barrier to the successful use of solar power, provided solar is used in a way that conforms to the strengths of the technology.

But bad weather can also occur during periods of peak demand, resulting in diminished output from a solar plant. With reliable weather forecasting, operators of a solar plant can reduce their uncertainty about whether their plant can produce electricity on a certain day, and armed with a good weather report, they can bid for short-term peak load power contracts with some confidence. (Although even a cloud drifting between the Sun and the plant can cause plant output to plummet.) But even modest confidence about the ability of a solar plant to manufacture power on a given date diminishes the further into the future the date is chosen. As a consequence, solar generating units introduce instability into the supply of electricity. One way of compensating for this instability is to depend on natural gas–fired power plants to make up the shortfall on those days when solar units cannot produce. Some solar units are even built next to natural gas–fired plants for this very reason. (This was done, for example, with the SEGS parabolic trough power plants in Kramer Junction, California.) Under these conditions, of course, it is the natural gas–fired unit, not the solar unit, that guarantees system reliability. From the point of view of system reliability, therefore, it might be more accurate to describe these generating stations as natural gas–fired stations supplemented by solar-powered units. The solar unit needs the natural gas–fired unit to ensure system reliability, but the natural gas–fired unit does not need the solar unit to accomplish the same goal.

There is another strategy besides building back-up fossil fuel plants to increase system reliability while at the same time incorporating solar power plants. The power supply can be made more reliable by building additional solar (or other renewable) generating stations in different locations to serve the same market. If the locations are far enough apart, the weather conditions at one location will not affect the weather conditions at other locations, and system reliability will be improved. To see how this works, imagine two widely separated locations, *A* and *B,* each with a solar plant. Under these conditions, if the probability that the weather will be unfavorable for power production at location *A* is 10 percent, and the probability that the weather will be unfavorable for power production at location *B* is also 10 percent, then the chance that the weather will be simultaneously unfavorable at *both* locations *A* and *B* is only 1 percent (0.1 × 0.1 = 0.01).

The reliability of this hypothetical two-unit solar power system is easy and instructive to predict under the assumption that they are far apart. Again, suppose that each location has a 90 percent chance of favorable weather and thus only a 10 percent chance of unfavorable weather. When the weather conditions at one site have no effect on the weather conditions at the other site, then the probability that it will be clear at both sites is just the probability that it will be clear at one site *times* the probability that it will clear at the other site. In other words, the probability that it will be simultaneously clear at both sites is 81 percent ($0.9 \times 0.9 = 0.81$). Finally, the probability that it will be cloudy at one site and clear at the other is 18 percent. This was obtained by subtracting the sum of 81 percent, the probability that it is clear at both sites, and the 1 percent probability that it will be cloudy at both sites, from 100 percent: ($100 - (81 + 1) = 18$).

Notice that this hypothetical system will fail completely only 1 percent of the time, but to obtain this higher rate of reliability two units had to be built, a substantial increase in costs. Furthermore, the two-unit system will produce full power only 81 percent of the time—that is, it works at full power only four out of five days on average. This is true even though the individual plants will operate at full power 90 percent of the time. The system will operate at half power 18 percent of the time, which is approximately one out of every five days. By building additional units in widely separated locations, system reliability can be increased, but at a high cost. Nor are costs limited to construction. Additional costs are incurred *because* the power system is distributed over a wide area. Improved connectivity will, in general, require the construction of long corridors of high-voltage transmission lines, which are themselves expensive and entail additional environmental costs.

Practically speaking, the preceding two-unit example is too narrow. There is, of course, no reason to restrict oneself to two solar-powered units or to solar-powered units at all. If the goal is to power as much of the system as possible using only renewable sources of energy, the same ideas that applied to solar-powered units can, for example, also be applied to wind-powered units and wave energy producers. Making a power system reliable while relying solely upon intermittent sources of energy is possible but expensive, and there is little to be done about it. Solar power plants require a level

Connecting Electricity Producers with Consumers: A Case Study

Between the electricity consumer and the electricity producer is a complex system of power lines. Some of the lines carry low-voltage electricity for local distribution, and some carry high-voltage electricity for transmission across longer distances. (Increasing the voltage for long-distance transmission reduces losses that result from the conversion of electrical energy into thermal energy as the electricity courses through the power line.) High-voltage transmission lines are expensive, highly specialized engineering projects. One of the challenges of renewable-energy technology—and solar energy, in particular—is that some of the best places to generate electricity from renewable sources are located far from the markets where the power is needed. In order to make use of this energy, it must be transported along high-voltage power lines.

In California, the San Diego Gas and Electric Company is currently planning to build a $1.3 billion high-voltage transmission corridor to bring power from geothermal power plants and solar-powered generating units in the Imperial Valley to San Diego County. The distance the electricity must travel is about 150 miles (240 km). The high-voltage lines will be suspended from enormous towers that opponents to the project describe as eyesores.

This high-voltage transmission corridor project has generated opposition from those who object to placing power lines through the Anza-Borrego Desert State Park. Some object to the impact the lines will have on the park, and others object to the project on the basis that, if approved, it will become the first of many high-voltage transmission corridors through state parkland. Others who live along the proposed path also object to 150-foot (46-m) transmission towers near their homes. But to be viable, large-scale solar and geothermal facilities must be built where the resources upon which they depend are located. In particular, solar plants must be built where land is available, where the weather conditions are optimal, and where the air is as clear as possible.

(continues)

(continued)

The California Independent System Operator, the authority responsible for routing power through the region's high-voltage transmission
network, asserts that the proposed transmission line is the most efficient way of meeting the region's power needs, but all parties agree
that it is not the only way. An older, less efficient natural gas–fired
power plant is located in the metropolitan San Diego area, and it could
be replaced with a high-efficiency natural gas plant. No new transmission corridors would be needed to connect the proposed replacement
plant with the market it would serve. "The most effective way to serve
electric demand is to locate the supply close to the load," Ali Amirali
is quoted as saying in a June 4, 2007, article in the *North County Times*.
(Mr. Amirali is managing director for regulatory policy and transmission
at Dynegy, the owner of the old plant.) And Mr. Amirali is right. But
generating stations are notoriously long-lived. A new natural gas–fired
plant would mean decades of additional carbon dioxide emissions and
continued uncertainty about electric power rates due to volatility in
the price of natural gas. By contrast, the solar plant is emissions-free,
and it will provide power at a predictable price, but it cannot be sited
in San Diego.

A third alternative is to build the corridor along a different route. But
while a new, and presumably more expensive, path would mollify the current set of critics, it would create new opposition among those who live
along the alternate route.

This controversy will almost certainly be repeated elsewhere as
power producers seek to build large solar plants in locations where solar
output is maximized and plant costs are minimized. If they are unable to
site their plants in locations with the best characteristics, their costs will
increase even as the amount of power they produce (and so the amount
of profit they can earn) diminishes. As profits decrease so will interest in
solar. These controversies call into question what it means to preserve the
integrity of the natural environment. How these disputes will affect the
future of solar energy is not yet clear.

of redundancy that more conventional systems do not, and these redundant systems are expensive. Currently, however, the economic costs associated with reliance on solar energy are easy to overlook because solar-powered generating stations contribute so little to the total power supply. Any additional costs incurred by the substitution of natural gas for solar energy, for example, are small because the amount of power being replaced is small relative to the size of the market. The same is true of using other renewable sources for backup. The economic costs of solar will draw increased attention if its market share grows. Solar energy is free, but the systems required to harness it are not.

AVAILABLE ENERGY

Planet Earth receives a tremendous amount of energy from the Sun, but the characteristics of this energy make it difficult to exploit economically. Some of the challenges have already been described: At Earth's surface, solar energy is intermittent, and its intensity is not entirely predictable; the intensity of solar energy depends upon latitude, and it is often most intense in places that are far from major markets and also far from high-voltage transmission lines. But these facts do not entirely convey the challenges involved.

Because the most convenient form of energy for transmission and use is electrical energy, a great deal of research continues to be devoted to finding better ways to convert solar energy into electrical energy. But the conversion of solar energy into electrical energy always involves losses. To say that Earth's surface receives 1,000 watts per square meter of solar radiation—which is true, provided the skies are clear and the piece of surface under consideration is near the equator and the Sun is high in the sky—provides little useful information about the relationship between land use and manufactured power. With an energy input of 1,000 watts per square meter of land (for part of a day), how much electrical energy can be produced per day per square meter of land? It is important to identify the relationship between land use and electricity production.

Parabolic trough power plant, located 35 miles (56.3 km) southeast of Las Vegas, Nevada

In this section, only concentrated solar power (CSP) systems will be considered because CSP technology currently makes by far the biggest solar contribution to the grid. Similar ideas can, however, be used to describe the relationships between photovoltaic (PV) systems and land use.

Recall that in CSP systems, solar energy is first converted into thermal energy and then the thermal energy is converted into electrical energy. At each step in the process there are losses in the sense that some of the energy used as input at a particular step is not converted into output energy. Consider, first, the problem of converting solar energy to thermal energy. On a windy day, for example, a CSP system will not operate as efficiently as it would operate on a windless day. As wind blows along the pipes that carry the primary fluid, it cools the fluid, so some of the thermal energy that was absorbed by the fluid will be transferred to the air before it reaches the heat exchanger. This energy is lost; it cannot be converted into electricity. Also, because reflectors are large and lightweight, wind can introduce vibrations and deformations in their shape, which leads to diminished reflector function. In fact, in a strong wind, operators of some CSP systems will shut them down to avoid damage to the reflectors.

Another source of loss is poor alignment of the reflectors that every CSP facility uses to concentrate the Sun's rays. When out of alignment, these reflectors, whether they are the heliostats at a power tower facility, the parabolic reflectors in a dish/engine system, or the troughs at parabolic trough facilities, fail to concentrate the Sun's energy to the fullest extent possible. Or the reflectors may be dirty, in which case they reflect less sunlight than they were designed to do. There are a great many other factors that contribute to losses during the process of converting solar energy to thermal energy, and their effects are generally cumulative.

Once some of the solar energy has been converted into thermal energy, the thermal energy must be converted into electricity. This process, common to all heat engines, is notoriously inefficient. First, it is important to understand that there is an upper limit to the percentage of thermal energy that can be converted into electricity. This upper limit is a law of nature; it cannot be exceeded, and the percentage of thermal energy that a CSP facility can convert into electrical energy is not very high. Of course, it is easy to create an engine that converts less thermal energy into electricity than is specified in this upper limit, but it is impossible to create an engine that converts a higher percentage of the heat into electricity. This percentage can be computed using only the temperature difference between the working temperature of the engine and the temperature of the environment. The smaller the temperature difference, the greater the losses, which is another way of saying that as the temperature difference between the inside of the engine and the outside diminishes, the larger the percentage of the thermal energy that cannot be converted into electricity. It is, in effect, wasted, and there is nothing to be done about this waste. Its existence is a law of nature.

The U.S. Department of Energy's National Renewable Energy Lab is the nation's premier laboratory for research into renewable energy. *(National Renewable Energy Lab)*

is, in part, dependent on these laws and regulations. This chapter describes some of the ways that governments have attempted to support solar power.

SOLAR ENERGY POLICY IN THE UNITED STATES

From the 1950s through to the early 70s, most of the interest in solar energy stemmed from the value of photovoltaic technology to the space program. NASA was the main customer for two reasons: (1) PV technology was especially well-suited to space, and (2) NASA was willing to pay the extremely high costs of these early solar cells.

There had been a number of small but vibrant solar energy markets prior to World War II. CSP technology and solar water heaters experienced commercial success without government support in some areas of the United States that had not yet gained access to inexpensive fossil fuel–based energy. But by the conclusion of World War II, fossil fuels had become ubiquitous and inexpensive. No other technology could compete with them, and in particular, solar technologies were driven into obscurity.

The situation changed again during the 1970s. First, in 1973, members of the Organization of Petroleum Exporting Countries (OPEC) sharply raised prices for oil, and for a brief period these nations placed an embargo on shipments of oil to the United States and the Netherlands for their support of Israel during its 1973 war with its neighbors. The effects of the price hike shocked many Americans. Long lines formed at gas stations; electricity produced by the many oil-fired power plants in operation at the time became very expensive, and the cost of living increased along with the price of oil. The situation was further exacerbated in 1979, when a revolution in Iran, long one of the world's biggest oil producers, caused a brief halt to oil production in that country. Again, oil prices increased sharply. It was during the 1970s that the United States government first attempted to craft policies that would speed the development of energy sources other than fossil fuels. One of the targets for rapid development was solar energy.

Since the 1970s, support for solar energy in the United States has, broadly speaking, taken a three-pronged approach. First, government research programs have sought to increase the economic competitiveness and reliability of solar energy technologies. Second, various financial incentives, often in the form of reduced taxes, have been offered to prospective buyers of solar technologies in an attempt to increase demand for solar powered equipment. Third, subsidies of various types have been offered to manufacturers of solar power products in order to increase supply. It is a comprehensive approach, but it has not always been successful. In the end, investing in solar products is still a choice, and that choice is influenced by a variety of factors, some of which have little to do with solar energy. The prices of alternatives, for example, most of which also receive government subsidies, always affect how potential customers behave. It is not enough for solar energy to be reasonably inexpensive or reasonably reliable; it must also be at least as inexpensive and at least as reliable as the alternatives. This is most easily seen in the failure of the United States to create a viable solar energy market during the years following the energy crises of the 1970s.

During the 1970s, the United States government established research programs in solar energy with the goal of developing technologies that would lessen the nation's dependence on imported oil. In addition to research, the Energy Tax Act of 1978 established subsidies in the form of tax credits for residential solar installations and other credits for business investments. A third method of support for alternative energy strategies was established in the Public Utilities Regulatory Policies Act of 1978 (PURPA). This law sought to encourage the development of renewable energy sources by guaranteeing nonutility energy producers a market. This was accomplished in the following way: PURPA required utilities to purchase electricity from nonutility power producers as long as the power was generated using renewable energy sources, a category that included, but was not limited to, solar. The price of the electricity was to be determined by the "avoided cost," or the amount of money that the utilities saved by purchasing the power rather than generating the power using their own facilities.

But government efforts of the late 1970s had little permanent effect because oil prices did not remain high for long. The very high oil prices that OPEC demanded also made it profitable for oil companies to develop other oil fields that would have been too expensive to develop when oil had cost much less per barrel. As new oil fields were developed, new oil began to flow onto the market, sometimes from non-OPEC countries, a situation that eroded OPEC's market share and drove down oil prices. The national perception of an energy crisis faded, and as alarm at the United States's vulnerability to hostile oil-producing nations dissipated, so did interest in solar energy. Government research programs were scaled back, and the residential credits program expired in 1985. Business credits were repeatedly renewed, however, and in 1992, they were extended indefinitely, but with only a very limited effect on business investment in solar technology. PURPA's effect on the development of nonutility renewable energy capacity was likewise very modest. In 1998, after 20 years of encouragement, solar, wind, and geothermal sources combined constituted just 3.7 percent of nonutility generating capacity, a small percentage of one small segment of the power-generating industry. During this period, the solar industry could not be subsidized to success.

Beginning in the late 1990s, three factors generated renewed interest in alternatives to conventional power plants: (1) the laws and regulations governing the electricity-generating business were substantially changed, (2) the price of natural gas increased sharply, and (3) there was increased concern about the effects of carbon dioxide emissions on the global climate.

In 1996, the federal government began a fundamental restructuring of the way that the electricity markets operated. Since the beginning of the electric power business, utilities generated the power that they sold; they transported that power along high-voltage transmission lines that they owned, and they distributed it through their own local networks. The utilities, which were government-regulated monopolies, owned and operated all aspects of their business. The only challenge to this system occurred with the introduction of PURPA, but PURPA was too limited in extent to change the structure of the industry.

In response to the 1996 federal legislation, the Federal Energy Regulatory Commission (FERC) issued two important regulations, orders 888 and 889, which restructured the rules by which electricity markets operated. These orders required the owners of the high-voltage transmission lines that connected power producers to power consumers to grant nondiscriminatory access to all power producers—that is, any power producer could use the network to transport electricity provided it could find customers to buy its

(power) Distributed Generation

For a long time, power producers built increasingly larger, more powerful generating stations to meet demand. There are several good reasons for the emphasis on large centralized stations. Large plants benefit from economies of scale—that is, the cost of power per kilowatt-hour is, as a general rule, cheaper when the kilowatt-hour is produced by a large plant than by a small one. Controlling power output is also easier when there are only a few plants producing power.

There are also disadvantages to centralized power production. Large plants have become increasingly hard to site. The task can be impossible if the power producer wants to site a plant near the market it serves. But siting a plant far from its intended market may mean that a high-voltage transmission corridor must be built. High-voltage transmission corridors have also become increasingly hard to site. Whatever its advantages and disadvantages, for many years, centralized power production was the only method of power generation because it was the only system that was economically and technically feasible to implement.

Today, a second model for power generation has captured the imagination of some engineers and activists. Called *distributed generation,* this method envisages a large number of relatively small power producers, each of which "injects" power into a local network or mini-grid. There are real advantages to distributed networks. Building multiple small power stations near their intended markets may be easier than building a few large ones, and because the stations are small, the supply of electricity can be increased in small steps. (Increasing capacity in small steps reduces the product. The hope was that by opening the power manufacturing aspect of the business to all interested parties, those that produced the lowest-cost power would succeed, and higher-cost producers would be forced to adapt or be driven from the market. (Whether this has succeeded is a matter of some dispute. As witnessed by the California energy shortages and electricity price increases of 2000,

risk to investors, who are always leery of building a single very expensive facility that might, when complete, produce far more power than there is demand at the time of its completion.) Another advantage to using a distributed system is that there are fewer losses due to the transmission of power across long distances.

Distributed power systems have only recently attracted interest because until recently the technology needed to implement a distributed power system was not available. Keep in mind that the production of electricity is a technically complex enterprise. Because there is no way of storing electricity, it must be produced simultaneously with demand. Controlling the output of a large number of small producers is technically more difficult not just in terms of matching supply with demand but also in terms of ensuing quality of supply—avoiding power surges, voltage drops, etc. Safety is an issue as well: Shutting down a section of the system for maintenance or in the event of an emergency is a more complex operation when many small power producers are operating almost independently of one another.

As efficiencies increase and costs decrease, solar power of all kinds can be expected to play a more important role in areas where distributed generation models are adopted. But it is not yet clear that the distributed generation model will be widely adopted. Whether its advantages outweigh its disadvantages remains to be seen. All that is certain is that distributed generation will continue to attract interest as power-production technologies and control technologies continue to evolve.

which occurred two years after California restructured its power industry, the idea has not been an unqualified success.)

After the restructuring, there was a boom in natural gas–fired plant construction, because these plants were comparatively easy to site and inexpensive to build. Initially, the fuel was also inexpensive, but domestic natural gas supplies, already tight, were further stressed by the additional demand caused by these plants. In response, prices began to rise rapidly. Hurricanes, especially Hurricane Katrina in 2005, disrupted supplies of natural gas from the Gulf of Mexico, one of the nation's richest sources of natural gas, and further increased costs. More importantly, perhaps, the hurricanes exposed the fragility of the nation's natural gas supply system.

Unable to meet domestic demand for natural gas with domestic supplies, expensive liquefied natural gas imports began to increase during this time, as did imports by pipeline, principally from Canada. The natural gas–fired power plant building boom began to fade in the face of high fuel costs, and interest in alternatives began to increase again. In theory, if entrepreneurs could build a solar plant that produced electricity at a competitive price, they would find an eager market for their product.

As already mentioned, concern about the cumulative effects of the emission of vast amounts of carbon dioxide into the atmosphere as a result of fossil fuel consumption also began to increase during the late 1990s. Some buyers of electric power voluntarily adopted standards requiring that a certain amount of the power they purchased be produced from renewable sources. In other cases, there were state mandates that required that a certain percentage of all power be produced from renewable sources. These are called portfolio standards. Efforts to incorporate mandatory portfolio standards into the federal 2005 Energy Policy Act failed, but production credits, which are production subsidies for facilities that use a variety of government approved technologies, solar included, did make it into the 2005 bill. Finally, the Energy Policy Act of 2005 provided tax credits aimed at reducing the cost of building solar power plants. Solar energy technologies had once again become heavily subsidized. As a result, new solar energy plants are now being brought online; others are under consideration, and the solar industry is again undergoing a period of subsidized growth. The costs associated with these programs are currently small, because the solar power industry is itself still small. As the industry grows, however, the costs of the subsidies will grow as well. It is not clear how long the nation will continue to subsidize solar energy at the current rate, nor is it clear whether the industry can survive without substantial subsidies.

SOLAR ENERGY IN GERMANY

The solar energy industry in Germany is bigger and more dynamic than the one in the United States, and so are the subsidies. But the way that solar energy is encouraged in Germany is also different from that in the United States. The German scheme is designed to encourage immediate investment in solar energy technology.

In Germany, the distinction between solar power consumers and producers is somewhat blurred by the method used to encourage the deployment of solar power technology. At present the German government guarantees a fixed price for energy produced by solar arrays and certain other qualifying renewable technologies, and the price is good for 20 years. That price depends on the size and type of plant. The goal of the subsidy is, of course, to reduce investor risk, thereby making investment in new solar facilities more attractive.

Energy investments are risky because the energy sector seems poised for a period of rapid change. The introduction of a less expensive technology could put economically marginal producers of electricity out of business. No matter how efficiently a solar energy plant (or any other plant) is operated, an innovation in another technology

The Fraunhofer Institute for Solar Energy Systems, one of the world's premier solar energy research institutions *(The Fraunhofer Institute for Solar Energy Systems)*

could make these plants obsolete. Indeed, this type of competition was the hope of those who restructured electricity markets in the United States. But in Germany, because the price paid for electricity produced by a qualifying solar facility is guaranteed for 20 years, it does not matter what happens to the market as a whole. The producer can, with reasonable certainty, calculate a minimum profit to be made from any given solar facility prior to investing in it.

To speed the development of the industry, the 20-year fixed price paid for energy produced by qualifying solar installations is reduced by 5 percent annually. This is one provision of Germany's 2004 Renewable Energy Sources Act. The goal is clear: The sooner a solar-powered facility begins producing electricity, the earlier it can lock in the price for the electricity it produces, and the more profit can be made.

Nor is this program restricted to commercial units. Residential solar units can also qualify. Germany is, therefore, on the way to developing a distributed system of power generation, in which power flows into and out of the grid from numerous smaller units. Other nations are also experimenting with this idea, but Germany is a world leader in implementing the concept. Whether distributed generation proves better or even as good as more conventional systems, which depend on big centralized power stations, is not known, because only in recent years has the technology for controlling a distributed system become available. These systems are works in progress. And even in Germany, most energy is still produced by large conventional power plants. The value of a distributed system may, however, first be proved or disproved in Germany, where approximately 1 percent, or 400,000 houses, have already been outfitted with solar panels, and that number is rapidly growing.

A quick look at a map will show that Germany is not well-situated to rely on solar power. Because it lies in northerly latitudes, winter days, when demand is high, are both cold and short, and some parts of Germany are famously cloudy during the winter. (The solar panels described in the preceding paragraph are deployed more in the sunnier south of the country than in the north.) Consequently, the contribution of solar energy to the nation will be modest for as long as the collectors are deployed within the borders of the Germany. The solution, for those interested in radically expanding the contribution of solar power to the German grid, is to look outside the country.

Across the Mediterranean Sea in North Africa is the Sahara. The northern Sahara is sunny, near the equator, and has a low population density. It is an ideal place to build many large solar power generating units. This proposal is receiving increasing amounts of attention in Europe and especially in Germany. The solar production facilities would be linked with consumers in the north by laying high-voltage power lines beneath the Mediterranean. Creating such an extensive high-voltage transmission network is an enormous expense, and Germany has sought financial support from its neighbors but has received a mixed reception. In Great Britain, in particular, the argument is that the system will not improve energy security because consumers will, essentially, be importing sunshine from the same politically unstable regions from which they currently import oil. The idea has, however, attracted support from some potential producers as well as some potential consumers.

To make the system worth the expense, it would have to be built on a huge scale and include facilities for thermal storage. The power facilities, provided they are built close enough to the Mediterranean, could also produce desalinated water as a co-product of electricity generation. North Africa is a dry region with a fast-growing population. The advantages to both parties are clear. On the producer side, there is the hope that, in addition to the water, a new and important industry will grow in a region where unemployment is high. On the consumer side is the hope for a new, abundant, and pollution-free source of energy.

This is a time of rapid change in both the economics and technology of power production. New technologies and new business models are being deployed. Solar energy currently has a place in this new market, but it is not clear whether it will grow significantly, maintain market share, or disappear again. (In order to increase market share it must grow faster than other competing technologies.) Predictions are difficult because despite the emphasis in some countries on solar energy, solar still fails to be competitive with many conventional technologies in several key areas. It is important to remember, for example, that even as Germany and Algeria negotiate about building the huge solar facilities described in the preceding paragraphs, Germany is engaged in a coal-fired power plant building boom.

Demand for electricity will be met. This much is certain. Only the technology used to produce the electricity is in question. Any nation that meets the demand for electricity with relatively high-cost methods of production (when low-cost methods are available) is choosing to spend its money on the *way* that electricity is generated rather than on the electricity itself. This diverts money from other, possibly more popular, programs. It is not clear how long in a democracy such a policy can be maintained. Engineering considerations as well as political, economic, and environmental factors will all help to determine the future of solar energy. It is a future that is far from clear.

A Brief History of Geothermal Energy

Geothermal energy means thermal energy from Earth's interior. It is also a term for a collection of technologies used to harness Earth's thermal energy for useful purposes. The term *useful purposes* often means converting thermal energy into electrical energy, but some engineers have also found ways to use Earth's heat directly to heat buildings, bathwater, or just to melt snow on walkways. Often, even the simplest-sounding application is based on sophisticated ideas about geology, the nature of heat, and the physical principles that govern the movement of heat from one place to another. To attempt to harness Earth's heat without some understanding of these ideas is to risk damaging or destroying a valuable thermal resource, and there are instances where this has already occurred.

Widely divergent claims have been made about the potential contribution that geothermal energy can make to the world's

Kilauea Volcano, Hawaii. Volcanoes demonstrate that there is a great deal of thermal energy within the planet. *(Hawaii Volcanoes National Park)*

energy needs. Higher estimates emphasize that beneath Earth's crust, there is an enormous supply of thermal energy—from a practical point of view, an inexhaustible supply of thermal energy. Lower estimates emphasize that most of this energy cannot be accessed with today's technology. At present, both points of view are correct, and the future of geothermal energy will depend on technical breakthroughs within the field and in other areas of energy research. (Producers of geothermal energy must compete against other energy technologies.)

Although geothermal energy is often described as a renewable resource, it is not renewable in the sense that solar energy is renew-

PART II

Geothermal Energy

able. An operator of a geothermal power plant, in particular, must be very careful to monitor the effects that plant operation has on the geothermal energy supply or run the risk of permanently ruining the site. (By contrast, this cannot happen with solar energy.) Geothermal energy technologies must, therefore, be managed. They can be operated for long periods of time, sometimes indefinitely, but only if they are operated in a way that does not disrupt the geothermal resource.

This chapter begins by describing some 19th-century discoveries about Earth's age and temperature, and concludes with a brief description of the concept of *plate tectonics,* one of the most important scientific discoveries of the 20th century and a key concept to understanding why available geothermal resources are not distributed evenly across the surface of the planet.

HEAT, GEOLOGY, AND THE AGE OF EARTH

Until the 19th century, most Western intellectuals believed that Earth was less than 10,000 years old. The Scottish scientist James Hutton (1726–97) presented the first concerted challenge to this belief when, in 1795, he published his two-volume series *Theory of the Earth.* Hutton had found a geological formation called an unconformity that, he believed, proved that the Earth was much, much older than 10,000 years. Hutton believed that his unconformity was a record of the following series of events:

1. sediments had gradually accumulated beneath the sea and turned to stone;
2. this formation of sedimentary rocks had been tilted and lifted to form mountains;
3. erosion had worn the mountains down;
4. the sea had covered the remains of the mountains;
5. a second layer of sediments had accumulated on top of the remains of the mountains;

6. the second layer had turned to stone; and
7. the second formation of sedimentary rocks had been tilted and displaced to form a second mountain range.

Hutton did not know how much time was required for this series of events to occur, but he knew that Hadrian's Wall, a wall built in Britain by the Romans, had lasted 2,000 years with little maintenance. To wear down a mountain, therefore, would require far more than 10,000 years. Famously, he wrote of his investigations:

> The result, therefore, of our present enquiry is, that we find no vestige of a beginning—no prospect of an end.

A great deal of energy is required to form a mountain range. Hutton believed that heat from deep within the planet provided the energy needed to create the mountains.

But Hutton failed to provide even an estimate for Earth's age. He had uncovered evidence that it was much older than 10,000 years, but his observations led him no further than to conclude that the planet's surface had experienced long slow cycles of uplift and erosion over and over again for as far back as he could observe.

The English naturalist Charles Darwin (1809–82), the author of *On the Origin of Species by Means of Natural Selection*, also believed that Earth must be much older than 10,000 years. In the *Origin of Species* Darwin had argued that species change slowly over time, and he had observed and documented enormous biological diversity. He deduced, therefore, that Earth had to be very old or there would not have been enough time for such diversity to have evolved. Darwin, an insightful geologist as well as a naturalist, had no way of transforming this insight into a logically coherent estimate of Earth's age.

The first coherent estimate of Earth's age was made by the Scottish engineer and physicist William Thomson (later Lord Kelvin) (1824–1907). Thomson knew something about the rate at which heat moved through rock, and based on the available observations he assumed that as one descended beneath Earth's surface, temperature increased an average of one degree Fahrenheit for each 50 feet (0.04°C/m) of depth. He further assumed that Earth began its existence as a ball of molten rock with an initial temperature of 7,000°F (3,900°C) throughout. He calculated that it would take anywhere from 24 million to 400 million years for the molten sphere to cool to its present temperature. The uncertainties in his prediction reflect uncertainties in his assumptions about the rate at which heat moves through rock and the manner in which temperature changes with depth.

Today it is known that Earth is a little more than 4.5 billion years old, more than 10 times as old as Thomson's maximum estimate. Although Thomson's computations were rigorous, they were inaccurate. There were many things about Earth's interior that he did not know—currents deep within the planet's interior, for example, transfer thermal energy between layers. Nor did he know that the supply of thermal energy within Earth's interior is continually supplemented by the decay of radioactive materials. What is important about the insights of Thomson and Hutton—at least for purposes of this volume—is that they understood that there is an interplay between the amount and distribution of thermal energy within the planet and other planetary features: Hutton believed that heat provided the motive force for mountain-building, and Thomson believed that the planet's temperature profile revealed the planet's age.

Today, scientists know that Earth's interior is extremely hot—thousands of degrees hot. Beneath the continents, the temperature is low enough to support human activities such as mining only within a few miles of the surface. Earth's searing interior is even closer to the surface beneath the oceans, where, away from the continental margins, Earth's solid surface layer, which is called the *lithosphere,* is about 60 miles (100 km) thick, and near volcanoes, no matter

Magma Chambers

A *magma chamber,* despite its name, is usually not a massive under-ground chamber filled with molten rock. Studies have shown that most magma chambers are complex structures consisting mostly of solid rock. Pores within the solid rock are filled with molten rock. Magma chambers near Earth's surface are valuable natural resources that can, in theory, be used to provide power and heat. The value of a magma chamber depends, in part, on its size, shape, and depth, and no two magma chambers are identical.

To obtain an impression of the amount of thermal energy contained in a magma chamber, consider the magma chamber located at Yellowstone National Park, which occupies parts of the states of Wyoming, Montana, and Idaho. Yellowstone has been volcanically active for millions of years. When it erupted 2.1 million years ago, it released a volume of magma 6,000 times greater than that released at Mount Saint Helens in the state of Washington in 1980. The current magma chamber underlies much of the park, and the thermal energy it releases is responsible for the many geysers for which Yellowstone is famous.

The temperature of the Yellowstone magma chamber is about 1,500°F (800°C). The chamber is thought to be about 10 percent liquid rock by volume and is 40 miles (60 km) long and 25 miles (40 km) wide. It is irregularly shaped. Measurements, which are very rough, suggest that the highest parts of the chamber are about four miles (6 km) beneath the surface. The total volume of the magma chamber is roughly 10,000 cubic miles (42,000 km^3). (Because geothermal electricity production in geyser fields at other locations removed enough heat and water from those sites to disrupt geyser activity, the geothermal energy at Yellowstone is protected by legislation written with the goal of preserving Yellowstone's famous geysers.)

Since the 1980s, the U.S. Department of Energy has investigated the possibility of drilling directly into magma. Magma sites in Hawaii and

California, for example, seem accessible, and hold the promise of enormous reserves of thermal energy. The idea is to create a well that extends directly into the magma chamber and then to inject water down the well. The resulting high-temperature, high-pressure steam would be produced at greater volumes and higher temperatures than the steam available from today's more conventional geothermal sites. As a consequence, a power plant operating off this higher-quality steam would be able to operate more efficiently—that is, not only could the power plant produce a larger supply of electricity per unit time, it would also produce more electricity per unit mass of steam. That, at least, is the theory.

As one might guess, there are technical difficulties to overcome when attempting to drill in such a hot environment. The position of the magma chamber must be determined prior to drilling, which sounds simple but is not. Identifying the position of the magma chamber is a three-dimensional problem. One must know not just its latitude and longitude but also its depth. Drilling is expensive—too expensive to simply drill and redrill until a well meets with success—and without a reasonably reliable prediction of the location of the chamber, one is unlikely to obtain financing from any backer, public or private. Additionally, the drill must be lubricated and cooled, and accomplishing these functions in such a hot environment is not easy. Less obviously, perhaps, magma generally contains large amounts of dissolved gasses under extremely high pressure. Drilling into the magma provides a path to the atmosphere for these gases, possibly resulting in an explosive release of pressure at the surface. There are other difficult problems to solve as well. Interest in mining the thermal energy contained in magma waxes and wanes, but magma as a source of thermal energy has continued to attract interest in the United States, Japan, and elsewhere. This technology is not expected to be commercially feasible anytime soon.

where they are located, great reservoirs of hot rock may exist just beneath the surface. These enormous volumes of hot rock represent an energy resource. The goal of geothermal engineers is, therefore, to convert some of Earth's inexhaustible supply of thermal energy into electricity or to make use of it in other ways.

LOCATION OF GEOTHERMAL SITES

A map of Earth's volcanoes will show that they are not distributed evenly across the Earth's surface. Most volcanoes are located along great arcs—the most famous of which, called the Pacific Ring of Fire, encircles much of the Pacific Ocean. The distribution of volcanoes is just one of many patterns that a careful study of maps of Earth's surface reveals.

The first such pattern to draw widespread attention is the close fit between the western coast of Africa and the eastern coast of South America. The German geophysicist and meteorologist Alfred Lothar Wegener (1880–1930) noticed that the coastlines of the two continents formed an almost perfect fit. Nor was the fit purely geometrical. He discovered mountain ranges, mineral deposits, and fossils on both sides of the Atlantic Ocean that seemed to correspond with each other. One of the best-known examples of a less obvious "fit" is the similarities that exist between the Scottish Highlands and the Appalachian Mountains of the eastern United States. Wegener's investigations eventually led him to conclude that the continents were in continual motion, slowly plowing through Earth's crust like massive ships through thin ice. His book, *The Origin of Continents and Oceans,* first published in 1915, initially attracted attention, but it was mostly in the form of scorn, and Wegener's work was eventually forgotten. It was the consensus view that what Wegener proposed could not occur.

In the 1960s, a rapid series of discoveries suggested a new theory called plate tectonics, a theory that was a modification of Wegener's. According to the theory of plate tectonics, Earth's crust, called the

Kilauea Volcano, Hawaii, 1954 eruption. The conversion of geothermal energy into electricity poses a unique set of challenges. (*J. P. Eaton. May 31, 1954, USGS*)

lithosphere, consists of about 12 distinct irregularly shaped "plates," some large and some small. These plates, which are fairly rigid, slide over a weaker, more plastic layer of rock called the asthenosphere, often scraping against one another or colliding with each other as they move. Sometimes when two plates collide, the edge of one plate is driven beneath the edge of another, forcing the lower plate to descend into the asthenosphere, where it eventually melts, causing the formation of volcanoes above the descending slab.

Although there are many independent lines of evidence that support the theory of plate tectonics, the beauty of the theory from the point of view of energy production is that it accounts for the existence of the vast arcs of volcanoes that stretch across Earth's surface. Most arcs form along the boundaries of tectonic plates.

But plates are not always colliding or scraping. In some places, two adjacent plates will move away from each other, and as they separate magma wells upward and fills the resulting rift in the

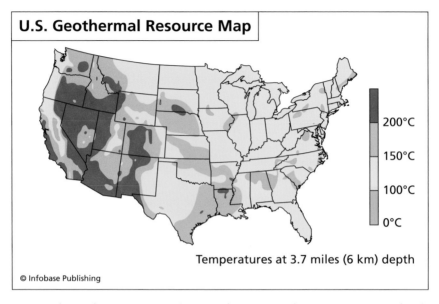

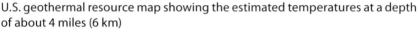

U.S. geothermal resource map showing the estimated temperatures at a depth of about 4 miles (6 km)

surface. The most famous example of this phenomenon, called the Mid-Atlantic Ridge, occurs along the middle of the Atlantic Ocean. The nation of Iceland is located along a segment of the Mid-Atlantic Ridge, and as a consequence, Iceland is very active volcanically. It is also a leader in the development of geothermal energy. More generally, there is a great deal of geothermal energy available along boundaries where plates are diverging.

Finally, there are a few volcanoes that are not located at the boundaries of plates. These are located above what are called *hot spots*. Each hot spot is a fountain of lava derived from a source located deep within the Earth. The fountain bursts through the plate above it to form a volcano. Because the plate above the hot spot is in continual motion and the hot spot is stationary, the resulting volcano is active only as long as it is above the hot spot. When the volcano moves off the hot spot it becomes extinct, and a new volcano begins to form. This can happen many times. The volcanoes form sequentially in time and in space. The most famous example of a hot spot is located in the Pacific Ocean. It formed that chain of volcanoes of which the state of Hawaii is but a part. All of the islands in the chain are volcanic in origin, but only the biggest, most easterly island—sometimes called the Big Island—is volcanically active, and it is very active indeed. Called Mauna Kea, it is presently located above the hot spot. This volcano is, at 33,000 feet (10,000 m), the tallest mountain on the planet. (Most of the mountain's bulk is hidden beneath the ocean.) The volcano of Kilauea, which is located on the Big Island, has been erupting continuously since 1983. Eventually, this island will also move off the hot spot, and its volcanoes will become extinct. The further west one goes in the Hawaiian chain, the older the island. The chain stretches 1,500 miles (2,400 km) from the hot spot. Hot spots are also rich sources of geothermal energy.

Most of the plates that compose the planet's surface are large, and hot spots are rare. Most cities are located in the interior of the plates, far from the boundaries, and farther still from any hot spots. Because the deposits of hot rock that are located near the surface are generally found near plate boundaries or near hot spots, readily accessible supplies of geothermal energy are generally located far from the energy markets. As a consequence, any power plants that made use of this energy would usually be located far from the markets that they would be built to serve. Bridging these distances with high-voltage transmission lines in order to bring the electricity to market is a separate and very difficult challenge. (See "Connecting Electricity Producers with Consumers: A Case Study.") Of course, large stores of thermal energy are also located directly beneath every city on Earth, but these are found at much greater depths than those along plate boundaries or near hot spots. Converting the thermal energy found in deep geological formations into electricity also presents formidable and as yet unsolved challenges. Research into geothermal energy continues, however, because the rewards may be worth the effort. Geothermal energy, properly managed, produces very little pollution, and it is highly reliable.

EARLY GEOTHERMAL TECHNOLOGY

Until the 19th century, hot springs were used for bathing or cooking and occasionally for heating nearby buildings. There was little remarkable or innovative in these applications. Provided hot springs were available, nearby residents recognized that it was easier and cheaper to use the preheated hot water that emerged from the Earth than to heat the water themselves. Romans, Native Americans, people throughout history and in many areas around the world made use of geothermal energy in these simplest of ways.

Ideas about geothermal energy began to change toward the end of the 19th century. So-called direct uses of geothermal energy, which involve using the thermal energy of the hot water without converting it into electricity, became more sophisticated. In 1892, the citizens of Boise, Idaho, installed the world's first geothermal district heating system. Water, heated by the Earth, was pumped through a series of pipes to heat a few buildings in downtown Boise. Within a few years the system expanded to heat approximately 200 homes and 40 businesses. That system, still in use, was for many years something of an anomaly even in Idaho. Built during a period of cheap fossil fuel energy, Boise's system was widely ignored. Interest grew, however, following the energy crises of the 1970s. During the 1980s, three more heating districts were added in Boise, and by that time several areas around the world were using hot water taken from the Earth to heat homes and businesses.

The problem with these early direct-use applications is that they are local in nature. The thermal energy is used near where it is produced because transporting thermal energy across long distances is difficult. Thermal energy is usually transported in water—that is, in order to transport heat, one must transport water or some other material. Water is preferred because it is inexpensive and can carry a great deal of heat without changing phase. It is also relatively easy to transport via pipelines. But because heat always flows from hot to cold, thermal energy is continually being lost as the hot water flows through the pipelines. When the pipelines are long, what emerges from the far end has cooled considerably. To make better use of geothermal energy, it needs to be converted into electricity, a form of energy that can be transported across much longer distances with less in the way of loss.

The technology needed to transport electricity long distances was developed, at least in a primitive form, during the late 19th and early 20th centuries in Europe and in the United States. The basic technology needed to convert thermal energy into electrical energy was also developed about this time. The processes by which thermal energy is converted into electrical energy were (and are) largely insensitive to the manner in which the thermal energy is produced. During the early years of the 20th century, coal, oil, and wood were, for example, all used as fuel to produce high-pressure steam that was then used to produce electricity. It was only a matter of time before someone thought of using geothermal energy to produce electricity. The first experiments in geothermal power production occurred in 1904, in Larderello, Italy, at a location where steam was erupting spontaneously from the Earth's surface. The existence of large quantities of hot, dry steam made the formation at Larderello relatively simple to exploit for power production. Essentially, one sunk a pipe into the formation, steam shot up the pipe, and the hardware necessary for electricity production was attached to the top of the pipe. The technology is simple-sounding, and much of it had, in fact, already been used at other power plants around the world—plants that used other sources, usually fossil fuels, as primary energy sources. What was new with respect to geothermal power production was that Earth functioned as the boiler. In practice, this meant that plant operators had to use steam with unusual chemical impurities. They were also faced with the problem of making optimal use of the geothermal resource without destroying it. The operating experience gained at Larderello was, however, of limited value because the type of geological formation found there proved to be very rare. The dry steam plant with the largest output is the Geysers, which is located in northern California. It was first harnessed for power production during the 1920s.

Early attempts to exploit the Geysers for power production occurred when competing forms of energy were relatively inexpensive. Consequently, the geothermal power produced at the Geysers proved to be economically uncompetitive, and production was soon shut down. Large-scale geothermal power production in the United States began in earnest in 1960, when an 11-MW plant began operation at the Geysers. Interest in geothermal energy intensified after the energy crises of 1973 and 1979.

In locations other than the Geysers and Larderello, Italy, wells for geothermal power production produce either hot liquid water or a combination of hot liquid water and steam. To convert the thermal energy at these sites into electricity other technologies were needed. (These technologies are described in chapter 10.) But as with the technology used at Larderello and the Geysers, these other power-production technologies had already been developed for other applications. To be sure, some modifications were necessary to adapt the power-production technology to geothermal applications, but as long as there was relatively easy access to the reservoirs of thermal energy beneath the surface, the modifications were evolutionary rather than revolutionary.

Today, a number of countries with easy-to-tap geothermal resources convert geothermal energy into electricity. And as with the early facilities, most of the technology in use at these sites was first developed for purposes other than geothermal power production. The technology needed to drill the wells that produce the hot fluids was developed by the oil and gas industry, and the equipment needed to convert the thermal energy into electricity is certainly familiar to power plant engineers at many other types of power-production facilities. The challenges still lie in adapting geothermal technology to the local conditions in order to make optimal use of the resource. It is also still true that the total amount of power produced by the industry remains small compared with demand—a situation that reflects the small number of available geothermal sites. But in response to demand for more power, engineers have begun to search for ways of tapping the much larger and much harder-to-access supplies of thermal energy that exist deep within Earth's interior. These efforts will require new ideas and new technologies. This part of the history of geothermal power production has yet to be written because it is only now being created. The challenges involved in accomplishing this new goal are formidable and are described in the next chapter.

or mineral. Uranium, for example, is a valuable element, and it is present throughout the environment. There are trace amounts of uranium everywhere: in soil, groundwater, in the oceans, in rocks, and even within human bones. Despite its ubiquity, in most locations there is not enough uranium to mine. In order to obtain usable amounts of uranium, it is, therefore, necessary to find more concentrated deposits of ore. The same is true of thermal energy, which is also located throughout the environment—in soil, groundwater, the oceans, rocks, and within our bodies—and in most locations there is not enough thermal energy to easily exploit. In order to efficiently convert thermal energy into electricity, it is best to find concentrated deposits of it and then "mine" it. Mining thermal energy is, of course, different from mining uranium ore. To mine uranium, it is necessary to remove large quantities of dirt and rock to access the ore. Much less matter needs to be removed to mine thermal energy, but as will be seen, some matter must also be removed in order to access this valuable resource.

The first step in siting a geothermal power plant is to find a concentrated deposit of thermal energy. At the present time, the best sites consist of hot, water-saturated, permeable rock, and for reasons that are explained later, the hotter the better. Because the Earth consists largely of hot rock, there is no question about the existence of adequate supplies. But these supplies are, for the most part, unavailable, either because they are too far beneath the surface or because the formations of hot rock are dry and impermeable.

Currently, a deposit of thermal energy has little commercial value if it is located more than three miles (5 km) beneath the surface. Volumes of rock with a temperature of 480°F (250°C) may be hot enough—that is, the thermal energy may be concentrated enough within the rock—to be worth exploiting.

Assuming that a deposit of thermal energy has been located at a depth that is both technically accessible and affordable, the next step is to bring the thermal energy up to where it can be converted into electrical energy. This is done with water. Essentially, the water absorbs the thermal energy at depth and then is raised to the surface. Although easy to describe, the process is rarely a simple one, and each locale must be analyzed separately. As has already been mentioned, only formations consisting of permeable, water-saturated, hot rock are commercially viable using today's technology. (*Permeable* means that water can move relatively easily through pores or fractures within the rock.) As might be expected, commercially viable deposits of geothermal energy depend upon a very special combination of circumstances, and sites that are located within drilling depth and that are hot enough, permeable enough, and also saturated with water are somewhat rare.

There are three types of commercially viable sites. They are classified by the characteristics of the water as it is extracted at the wellhead. If pure steam comes out of the wellhead, the site is classified as vapor-dominated. These tend to be the most profitable and the easiest to operate. If what comes out of the wellhead is a mixture of hot liquid water and water vapor, the system is called a hot-water system. Finally, if liquid water emerges from the wellhead that is "hot" but not very bubbly with steam, it is classed as a moderate temperature system. Moderate temperature systems can be used to generate electricity, but the procedure is more complicated than those used in vapor-dominated systems or in hot-water systems. They are also less efficient in their use of thermal energy and more expensive to build than the other two types of systems.

Having produced the thermal energy along with the water, the next step in the process is to convert the heat in the water into electricity. (The methods by which this is done—methods that are not unique to geothermal power stations—are described in the next chapter.) Having produced the electricity, the operator is now left with the problem of what to do with the water. Some early geothermal energy projects just dumped the water on the surface. This created two problems.

Geothermal Energy

Drilling

Power plant

Thermal reservoir

© Infobase Publishing

A geothermal system consists of the power plant, the thermal reservoir, and wells to connect the two.

First, hot water drawn up from these formations generally contains a variety of compounds dissolved within it, some of which can pose modest environmental problems. The composition of water drawn from these thermal deposits depends on the specific characteristics of the site. By way of example, sulfur dioxide, a chemical compound commonly found deep below the surface, dissolves easily in water to form acid. Simply drawing large volumes of mildly acidic water out of the ground and dumping them on the surface is, from an environmental viewpoint, irresponsible. Consequently, something else must be done with the wastewater. Second, the water that exists deep within a rock formation is finite in extent and can be withdrawn faster than it can be replenished. When this happens, the rate of flow from the formation diminishes and less thermal energy can be obtained because there is less water available to carry it to the surface. Withdrawing water too quickly from such a formation was a mistake made by operators of some early geothermal power stations. They damaged the very resource upon which they were dependent, and the result was a temporary or sometimes a permanent reduction in the stations' power output.

The solution to the problem of wastewater sounds relatively simple: Designers now drill a second well (or a set of wells) at some distance from the wells they use to withdraw hot fluids and use the second set of wells to inject some or all of the cooler "used" water (often with supplemental water supplies) back into the formation from which it was withdrawn—thereby recycling the water. Because the rock is permeable, and because the operators are continually withdrawing water from the formation, the injected water rapidly diffuses through the hot rock, where it is reheated. Over time, some of the injected water moves back to the point from which it was withdrawn, and the cycle begins again. The goal of engineers is to create a system in which as much water is injected as is removed. By balancing the rate of flow into and out of the formation, a geothermal power system can often be operated indefinitely and with little impact on the environment.

Although there are a number of commercial geothermal power plants in operation, their contribution to the total electricity supply is not large. Expanding the use of geothermal power will involve

(continued on page 120)

Fenton Hill, New Mexico: The First Enhanced Geothermal System

The first attempt to investigate the possibility of producing geothermal energy from hot, dry, impermeable rock took place at Fenton Hill, New Mexico. The idea for the project originated with a group of engineers and scientists at the Los Alamos National Laboratory, which is also located in New Mexico. In 1970, the Los Alamos team began to investigate a way to demonstrate the practicality of the concept of producing a reliable supply of thermal energy from a formation of hot, dry, impermeable rocks located deep underground. Their proposal, for which they received a patent, involved drilling into hot crystalline rock and then injecting water under pressure in order to create a vertical fracture. Once they had created the fracture, they planned to drill a second well to extract the heated water. Fenton Hill was a research project. The researchers never built a power plant. The goal was simply to demonstrate that they could create a closed system, simultaneously injecting cold water into one well and withdrawing hot water from the other.

The project, because it is simple to describe, sounds simple to accomplish, but it was both difficult and expensive. The first well, drilled in 1974, went down 1.8 miles (2.9 km), at which point they created the fracture. Intersecting the fracture with the second well proved challenging. Keep in mind that although they knew a good deal about the chemistry of the rock into which they were drilling, that information was not enough to predict the details of the fracture that they had created. In particular, they did not know the precise shape or location of the crack they had created almost two miles (3 km) beneath their feet. The first attempt to drill a borehole so as to intersect the fracture was unsuccessful. In fact, several attempts to create a second borehole linked to the first borehole through the fracture were unsuccessful.

They eventually created the sought-after closed system, and began a long sequence of measurements and tests in order to understand the physical characteristics of the resulting system. For example, during early tests the temperature of the water produced at the wellhead diminished with time—that is, the longer they pumped water through the thermal reservoir, the less heat the water absorbed. They concluded that the rock surface across which the water flowed was too small. As a consequence, they were removing heat from the rock adjacent to the fracture faster than it could be replenished by heat from more distant rock within the formation. In other words, their efforts had created a small volume of cool rock within a much larger volume of hot rock. The eventual solution was to create additional fractures through which the water could flow.

They also had to measure the volume of water produced at the wellhead (1.5 gallons [5.7 liters] per second), and the amount of pressure that had to be exerted at the injection site to maintain this rate of flow (82 atmospheres), and they measured changes in the chemical composition of the water after it had passed through the formation.

Later, the Los Alamos team drilled more wells at greater depths and created more fractures, each time gathering additional data about water flow, pressure, temperature, and chemical composition, and they gained additional experience with the technology involved. The Fenton Hill site was operated for about two decades before it was finally closed in 1994, and during that time other countries, notably Germany and Japan, contributed funds and personnel. (During the late 1980s sophisticated geothermal research programs were established in France and Japan.)

What the Los Alamos researchers did not do, however, was successfully demonstrate the commercial feasibility of their idea. Even now, several technical issues remain to be resolved. In particular, researchers must demonstrate that a large-scale Fenton Hill–like system can be operated for many years without significantly cooling the reservoir. Creating a large system would be expensive, so it would have to operate for a long time before it began to turn a profit. There is still a significant role for government research.

Krafla Geothermal Station in northeast Iceland. Despite having the richest geothermal resources in the world, in Iceland most electricity is generated with hydroelectric power. *(Mike Schiraldi)*

height increases the water pressure at the base of the dam so that when tunnels within the dam are opened and water flows through them, the water travels rapidly and with a great deal of force. As the water flows through a tunnel, it is directed against a turbine. A turbine is a device that converts the straight-line motion of the rushing water into rotary motion. In concept, a turbine is similar to a waterwheel, but is more sophisticated in design. The turbine is connected to the generator via a shaft so that as the turbine spins so does the generator. Electricity is the result. In brief: Rushing water is directed against the turbine, and the spinning turbine drives the generator.

As with hydroelectric facilities, power stations that convert thermal energy into electricity also have turbines, and as with a hydroelectric facility, the turbine at a thermal facility is connected to the generator by a shaft so that when the turbine spins so does the tests the temperature of the water produced at the wellhead diminished with time—that is, the longer they pumped water through the thermal reservoir, the less heat the water absorbed. They concluded that the rock surface across which the water flowed was too small. As a consequence, they were removing heat from the rock adjacent to the fracture faster than it could be replenished by heat from more distant rock within the formation. In other words, their efforts had created a small volume of cool rock within a much larger volume of hot rock. The eventual solution was to create additional fractures through which the water could flow.

They also had to measure the volume of water produced at the wellhead (1.5 gallons [5.7 liters] per second), and the amount of pressure that had to be exerted at the injection site to maintain this rate of flow (82 atmospheres), and they measured changes in the chemical composition of the water after it had passed through the formation.

Later, the Los Alamos team drilled more wells at greater depths and created more fractures, each time gathering additional data about water flow, pressure, temperature, and chemical composition, and they gained additional experience with the technology involved. The Fenton Hill site was operated for about two decades before it was finally closed in 1994, and during that time other countries, notably Germany and Japan, contributed funds and personnel. (During the late 1980s sophisticated geothermal research programs were established in France and Japan.)

What the Los Alamos researchers did not do, however, was successfully demonstrate the commercial feasibility of their idea. Even now, several technical issues remain to be resolved. In particular, researchers must demonstrate that a large-scale Fenton Hill–like system can be operated for many years without significantly cooling the reservoir. Creating a large system would be expensive, so it would have to operate for a long time before it began to turn a profit. There is still a significant role for government research.

(continued from page 117)

using a wider class of thermal reservoirs instead of depending solely on relatively shallow formations of hot, permeable, water-saturated rock. The first, and perhaps the most obvious, restriction to be removed is the restriction on depth. As technology improves, designers continue to drill deeper. But the maximum depth of a well is only partly a function of technology. It is also a function of economics. When the price of electricity goes up, the depth to which a company can afford to drill goes down. Conversely, cheap electricity means that drilling to greater depths becomes uneconomical.

Deeper drilling exposes more hot rock, but while the temperature of the rock is important, the permeability of rock is just as important. Permeability is crucial because water must be able to move freely through the rock formation in order to absorb the thermal energy present in the formation without significantly reducing the temperature of the "thermal reservoir." Ideally, the circulating water will absorb only a small amount of heat from each small volume of rock past which it moves. This small quantity of thermal energy can then be quickly replenished by the flow of heat from elsewhere within the thermal reservoir. When this occurs, heat can be removed from the thermal reservoir with little effect on its overall temperature, and the whole process can be continued indefinitely.

Unfortunately for the development of geothermal power systems, most deposits of concentrated thermal energy are found in formations of hot, dry, *im*permeable rock. A necessary condition for commercially exploiting these thermal reservoirs is to increase the permeability of the rock. To accomplish this, engineers must use a technology first developed by petroleum engineers. They must create fractures within the rock. The technique is simple to describe: First, they drill into the hot, impermeable rock, and then they force water down the resulting borehole at pressures high enough to fracture the rock formation below, thereby increasing its permeability. Whatever water is in the rock is now free to flow. If the rock is water-poor, as is often the case, an injection site must be created. This is accomplished by drilling a second well to a location within the thermal reservoir that is close enough to take advantage of the newly created fractures, but far enough to use as much of the hot rock for heating as possible. Once the second well has been created, water is forced down one well and drawn up through the second. Developing hot, dry, impermeable sites is where the future of geothermal power plant engineering lies. Such systems are called *enhanced geothermal systems* (EGS). There are no commercial EGS systems in operation today, but many believe that they will play an important role in electricity production in the future.

HEAT ENGINES

The general mechanism by which thermal energy is converted into electricity is shared by most large power plants. The conversion is accomplished through a short sequence of steps and is deceptively simple to describe. It is important to keep in mind, however, that the technology that has been created to produce electricity from thermal energy at a large scale has become very sophisticated as engineers have continued to devise larger and more efficient power plants to meet the ever-growing demand for electricity. Nevertheless, the basic concepts behind many of these thermal energy-to-electricity conversion schemes would have been familiar to 19th-century engineers and scientists.

As previously mentioned, the purpose of every power plant is to spin a generator, a device for converting rotary motion into electrical energy. Everything in the power plant is built to serve this goal. Generators are generally massive, and they must be spun continually and at a predictable speed. To see how this is accomplished in practice, first consider a hydroelectric power plant, which, although it is not a heat engine, is probably the conceptually simplest method of producing electricity. At a hydroelectric facility, the purpose of the dam is to raise the height of the water behind it. The increased

Krafla Geothermal Station in northeast Iceland. Despite having the richest geothermal resources in the world, in Iceland most electricity is generated with hydroelectric power. *(Mike Schiraldi)*

height increases the water pressure at the base of the dam so that when tunnels within the dam are opened and water flows through them, the water travels rapidly and with a great deal of force. As the water flows through a tunnel, it is directed against a turbine. A turbine is a device that converts the straight-line motion of the rushing water into rotary motion. In concept, a turbine is similar to a waterwheel, but is more sophisticated in design. The turbine is connected to the generator via a shaft so that as the turbine spins so does the generator. Electricity is the result. In brief: Rushing water is directed against the turbine, and the spinning turbine drives the generator.

As with hydroelectric facilities, power stations that convert thermal energy into electricity also have turbines, and as with a hydroelectric facility, the turbine at a thermal facility is connected to the generator by a shaft so that when the turbine spins so does the generator. The goal of the engineer, then, is to cause the heat source to spin the turbine. This is accomplished by introducing something called a "working fluid," which in practice usually means water, although sometimes other fluids are used.

Here is how a heat engine works: Heat is transferred from the heat source to the working fluid. As the working fluid absorbs the heat, it turns into a vapor, but because the vapor is confined within pipes, it is unable to expand. As a consequence, the pressure exerted by the vapor on the walls of the pipes through which it is flowing is very high. Pumps move the vapor through the pipes towards a valve. As the vapor flows through the valve it passes from a region of high pressure to a region of much lower pressure. Free of the confines of pipes and valves, the vapor expands rapidly and with great force. The valve directs the expanding vapor against the blades of the turbine, causing the turbine to spin. The turbine converts the straight-line motion of the expanding gas into rotary motion. As the turbine spins, it causes the generator to spin, and, once again, electricity is the result.

Once the vapor has passed the turbine it is usually directed to a device called a condenser, where it is cooled until it condenses into a liquid. Usually, the liquid is pumped back to the heat source, and the cycle begins again. To summarize: Thermal energy is transferred from the heat source to the working fluid; the working fluid undergoes a phase change from liquid to vapor; the vapor is used to drive a turbine; and the turbine spins the generator. Finally, the vapor condenses and some or all of it is pumped back toward the heat source.

All geothermal power stations work on some variation of the scheme just described. There are, in fact, three major variations on the sketch given in the preceding paragraphs. The method chosen for a particular site depends on the temperature of the water as it issues from the wellhead. Understanding the differences between the methods is important to understanding the limitations of each technology.

The Generating Station

There is, as the preceding chapter makes clear, a great deal of engineering and scientific know-how required in order to make use of geothermal energy. So far, however, only the technology needed to bring the heat to the surface and the technology needed to maintain the heat flow have been described. These are the aspects that are unique to geothermal power stations. The method by which thermal energy is converted into electricity has only been described in a very general way. But power generation is a technical subject. While generalities can be helpful, they fail to convey the very important technical issues that are key to understanding the value and limitations of geothermal technology—or, for that matter, any technology.

This chapter begins by describing three methods of converting geothermal energy into electricity. They are not interchangeable. The characteristics of a particular geothermal site dictate which technology should be used, and the choice of which technology to adopt reveals a good deal about the economic value of the site. To see how all of these considerations come into play, the chapter contains a conversation with an expert on one of the world's most productive geothermal power stations.

TECHNICAL CONSIDERATIONS

The great French engineer Nicolas-Léonard-Sadi Carnot (1796–1832) believed that the same physical laws that had successfully been used to describe water could also be used to describe heat. His investigations into the nature of heat were guided by this analogy between heat and water. He noticed, for example, that when a warm object and a cool object were brought into contact with each other, heat spontaneously "flowed" from the warmer object to the cooler one until both objects were at the same temperature. He compared this with water's well-known tendency to flow from higher elevations to lower ones—a phenomenon that is often described as water "seeking its own level."

Nicolas-Léonard-Sadi Carnot, one of the founders of the science of thermodynamics *(AIP Emilio Segrè Visual Archives)*

He also knew that the amount of work that could be performed by falling water depended very much on the distance that the water fell. A particular volume of water falling a large vertical distance could perform more work—turning a waterwheel, for example—than the same volume of water falling a short vertical distance. Water, he knew, only performed work when it was moving *between* levels. These observations led him to surmise that the same was true of heat, which, at the time, was called "caloric." Caloric was, after all, also used to perform work—in a steam engine, for example. He concluded that the larger the temperature difference between the boiler and the environment, the more work could be performed by a given amount of heat. He described the difference between the temperature of the boiler and the temperature of the environment as the "temperature drop," in analogy with the vertical distance water dropped when turning a waterwheel or some other mechanical device. Upon this analogy, he developed a theory of heat that served him very well. His ideas were published in his book, *Reflections on the Motive Power of Fire.* It is a peculiar fact that although heat is not a fluid, many of the predictions that were based on the analogy between heat and water are essentially correct.

With respect to the development of geothermal power plants, probably the most important characteristic of a site is the temperature drop, the difference in temperature between the environment and that of the liquid or vapor as it issues from the well. The larger the temperature drop, the more work that can be performed for each unit mass of water, either liquid or vapor. It is important to keep in mind, however, that these remarks only apply to the *maximum* efficiency of a power plant. It is always possible to build an inefficient geothermal power plant, but there are compelling reasons to avoid inefficiencies whenever possible. In particular, inefficient plants are more expensive to operate, and since the purpose of a plant is to earn a profit for its investors, a more-efficient plant is always better than a less-efficient plant, all other things being equal.

Efficiency is an especially important concept to bear in mind when evaluating geothermal power plants because the steam produced at geothermal fields is often less than half the temperature of steam produced at coal or nuclear plants. As a consequence, geothermal power plants are less efficient than more conventional power plants in the sense that much larger amounts of steam are required by geothermal power plants to produce the same amount of electricity as either coal or nuclear plants. All other things being equal, therefore, conventional power plants are to be preferred over geothermal power plants. But all other things are never equal, and there is a balance that must be struck between actual plant efficiencies, plant construction and operating costs, and the environmental impacts associated with any technology. These are complicated issues.

The simplest type of geothermal energy system to harness is called a vapor-dominated system. Generating stations operating at vapor-dominated sites are generally more efficient than other types of geothermal power plants. Vapor-dominated sites are also quite rare. In these systems, what issues from a well is dry steam—that is, steam is produced but no liquid water. Vapor-dominated systems can produce tremendous amounts of dry steam. To use this resource, pipes are connected to the wellhead to maintain pressure and to provide a path from the wellhead to the turbine. Steam, as it issues from the well, contains particles of dirt; these must be removed, and then, as described in the preceding chapter, the pressurized steam flows through a valve that directs it against the blades of the turbine. As the steam passes through the valve, it expands and drives the turbine, which drives the generator, and electricity flows into the grid. Once past the turbine, the steam is routed to a condenser, where it condenses into a liquid. The condenser, which causes a pressure drop downstream of the turbine, increases the efficiency of the energy conversion process and facilitates the separation of the water from any other gases that may be present. Finally, the liquid, or at least some of it, is injected back into the well to maintain the

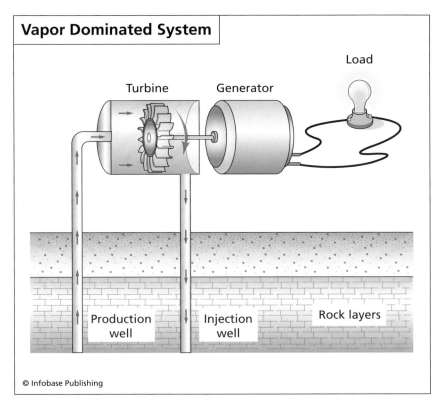

Vapor Dominated System

Load

Turbine   Generator

Production well   Injection well   Rock layers

© Infobase Publishing

Diagram of a power plant built to exploit a vapor-dominated system *(Source: EERE)*

supply of steam. Vapor-dominated systems tend to be mechanically simpler than other types of geothermal systems, and they are also more efficient in their use of geothermal energy, but there are not enough of them to satisfy the demand for geothermal power. The Geysers, operated by Calpine, is a vapor-dominated system.

Hot-water systems, which are more common than vapor-dominated systems, depend on thermal reservoirs that are saturated with liquid water rather than water vapor. These are sometimes called hot-water systems and sometimes called flash systems. The water is still very hot, sometimes exceeding twice the boiling point of water at atmospheric pressure (212°F [100°C]), but because the pressure deep beneath Earth's surface is so much higher than at the surface, the water remains in a liquid state. As the water flows up the well toward the surface, the pressure begins to drop, and the water begins to boil within the well, and some, but usually not all of it, turns to steam. What arrives at the wellhead is a mixture of hot liquid water and fairly high-pressure steam. The pressure on this mixture is reduced somewhat in a device called a "flash tank," which allows more of the water to change into steam. (It is to the operator's benefit to produce all of the pressurized steam possible.) The mixture of steam and liquid water, which is still under substantial pressure, is separated into the steam component, which is directed toward the turbine, and the

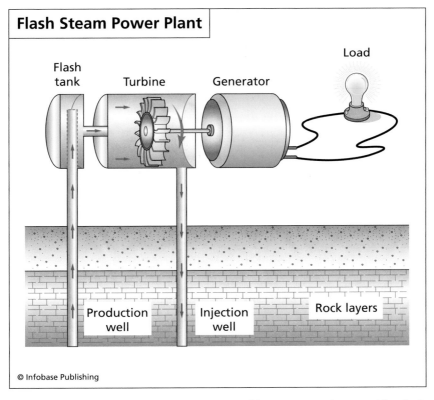

Flash Steam Power Plant

© Infobase Publishing

The production well yields a combination of hot water and steam. The flash tank maximizes the production of steam. *(Source: EERE)*

liquid water, which is pumped into the injection well. The steam now drives the turbine-generator pair in just the same way as in a vapor-dominated system, and once past the turbine, the steam is cooled in a condenser until it becomes a liquid. The water from the condenser, or at least some of it, is also injected back into the thermal reservoir, and the cycle is complete. This technology is employed, for example, at the Coso Hot Springs geothermal field in California, at facilities owned by CalEnergy Generation.

The third method of converting geothermal energy into electricity is used when what emerges from the production well is water that is hot, but not hot enough to form large amounts of steam. To be clear, the temperature of this water is still higher than the boiling point of water at atmospheric pressure, but a conventional power plant needs a great deal of steam delivered at a fairly high pressure to drive the turbine. These hot-but-not-quite-hot-enough systems are called moderate-temperature systems. The lower-temperature water in a moderate-temperature system is unsuitable for directly driving a steam turbine. It does not have enough energy. But liquid water has a fairly high specific heat (see chapter 5 for a discussion of specific heat), which means that even at a moderate temperature the water still carries significant amounts of thermal energy. The problem for engineers, then, is to find a way to make use of this thermal energy to create a relatively high-pressure vapor.

The solution to the moderately heated water problem is to use a so-called binary system, one that uses a working fluid other than water. Although most power plant designers choose water as a working fluid, many other choices are possible. Heat engines have been constructed that use ammonia, carbon dioxide, helium, or more exotic liquids and gases. Engineers tend to choose water because it is both abundant and inexpensive, but sometimes other considerations are more important. In geothermal power plant design the most important consideration is that the working fluid changes phase, from liquid to vapor, under the conditions prevailing at the plant.

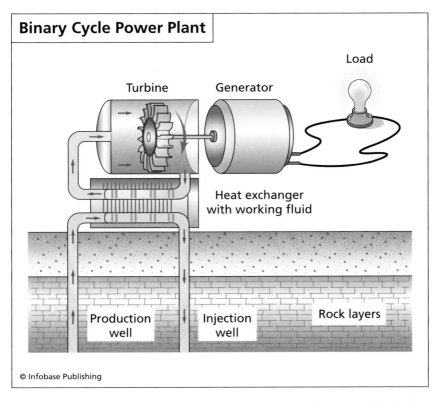

Binary Cycle Power Plant

Load

Turbine   Generator

Heat exchanger
with working fluid

Production
well

Injection
well

Rock layers

© Infobase Publishing

Binary cycle power plant. This system uses two separate loops, one for the hot
water and one for the working fluid that drives the turbine. *(Source: EERE)*

A binary system uses two "loops," each carrying a different
fluid. The first loop in a binary geothermal system contains the hot
water produced from the well. This fluid flows up the well to a heat
exchanger, where some of the thermal energy carried by the water
is transferred to the working fluid. After passing through the heat
exchanger, the somewhat cooler water is injected back into the rock
formation, where it is reheated. (Recall that a heat exchanger is a
radiator-like device that permits the exchange of heat but not mass
between two fluids—and the word "fluid" means anything that
flows, either liquid or gas.) On the other side of the heat exchanger

(continued on page 141)

An Interview with John Farison on the Challenges of Producing Power at the Geysers

J ohn Farison is director of process engineering for Calpine Corporation at the Geysers geothermal field in northern California. His career has focused on solving problems related to impurities in geothermal fluids, including scaling, corrosion, and emissions abatement. He has a bachelor of science degree in chemical engineering from Ohio State University. The following interview took place on February 7, 2008.

Q: The resource that you're tapping at the Geysers is permeable rock— permeated with steam. How large is this resource?

A: The Geysers is the largest dry steam reservoir in the world. The Geysers' steam reservoir covers about 40 square miles at the surface, as defined by where steam has been found at depth in over 700 wellbores that have been drilled into it. The top of the steam reservoir is about 4,000 feet below the surface, and wells average about 8,000 feet deep, with the deepest wells being over 12,000 feet deep. The reservoir rock is estimated to be about 8,000 feet thick over that 40-square-mile area. So the volume of the steam reservoir is estimated to be about 50 cubic miles.

John Farison (*John Farison*)

Q: How is the Geysers different from other geothermal resources?

A: There are many geothermal reservoirs in the world, but there are relatively few—four or five—that are called "dry steam" reservoirs. When you drill a well into these, they produce, just as the name says, dry steam. Whereas all of the other geothermal resources have hot brine/hot water, and some of them will, when the pressure is released on the fluid at depth, flow under their own pressure. The hot water flashes or boils to form steam, and you have to separate out the residual water before you can use the steam. Here at the Geysers, the steam comes out of the ground without any excess water, and we're able to use it directly. We pipe the steam overland to a centrally located power plant, where it drives a turbine and spins a generator.

Q: Are you currently tapping most of it?

A: Yes. There are some areas that may be underdeveloped, and we're currently working to move into those areas. But I'd say the majority of the field has been developed.

You might consider the Geysers the biggest heat exchanger in the world. With geothermal energy, the resource is the heat energy contained in the rock, and steam and hot water—the fluids that we produce out of a geothermal reservoir—are, you might say, the "working fluid" used to extract and transport that heat. There is a certain amount of fluid that is in the rock when you first tap into it, and then as you produce it, that amount of fluid is reduced, but you still have additional usable heat in the rock. So if we can put more water down to continue to tap that heat, we can increase the output of the reservoir over time. We've been producing electricity at the Geysers for 47 years now, and it's going to produce for another 40 years or so. It is really a huge heat exchanger, with steam wells tapping into it to produce the steam, and water injection wells distributed across the field to put water back in.

Q: Roughly, what volume of rock is needed to support one power plant?

A: It varies. Every geothermal resource is different in terms of how much fluid is produced per acre.

Q: Even within the Geysers?

(continues)

(continued)

A: Yes, even within the Geysers. There is a different flow rate from each of the wells, and they change over time. That's something that we monitor, and if we can figure out that there is some problem with the wellbore, we can go back in and repair it. That is something that we are doing right now. We're repairing some wells, and adding some additional injection wells to distribute where we put the water to get the best return of steam to mine that heat out of the rock.

Q: And when the steam emerges from the wellhead, what is the chemical makeup of the steam? Does it have to be cleaned up before it's put through the turbine?

A: Geothermal steam picks up some impurities. It has rock dust. There are also some trace gases. There's carbon dioxide, hydrogen sulfide, a bit of hydrogen and methane. Those are the main gases. Inside the power plant, we have to have vacuum systems to remove those gases from the condenser to make the power cycle work.

Q: They are not removed before the power is generated?

A: No, they are removed after the steam has expanded through the turbine. The steam expands through a "condensing turbine," which converts the heat energy of the steam into mechanical energy to spin the electrical generator. As the power is taken out of the steam, the steam condenses, and it exhausts into a condenser that is operated at vacuum. And the gases that are in the steam are not condensed when hit with cold water or when you put it into a condenser with cold tubes, and those gases would build up if you didn't have a vacuum system such as a steam ejector or a vacuum pump to continuously draw those gases out of the condenser.

Q: Oh? At that point the trace gases are removed and disposed of, and some of the water that has been condensed is injected back into the formation? Is that how it works?

A: The steam that has been condensed is at about 120°F. We send that water to a cooling tower, where a lot of the water is evaporated to cool the remainder. The remaining water is cooled back down to 75°F or 80°F and is sprayed back into the condenser or pumped through the shell and tube condenser to condense the incoming steam. So it's a cycle. Virtually all steam power plants operate the same way; whether it's fired with nuclear power or fossil fueled to boil the water, the power cycle works the same way. The main difference here is that our boiler is Mother Earth.

Q: I was under the impression that some of the condensed water recovered at the Geysers was injected back into the steam reservoir.

A: Yes, about 25 percent of the steam that is produced is recovered and injected as condensate back into the steam reservoir. The remaining 75 percent of the water that comes out of the ground is evaporated in the cooling towers. We now replace most of that lost volume of working fluid by injecting reclaimed water pumped to the Geysers from Lake County and Santa Rosa.

Q: With respect to the individual plants, these plants have been installed one after another over the years. Have the designs of the plants at the Geysers changed over the years?

A: There have been some design changes. The initial plants had "direct contact condensers" meaning that they had spray nozzles. Water was sprayed into this open chamber, hitting the steam as it exited the turbine. The steam would then condense, collapse, and form a vacuum. Over time we've moved to a new design that has a "shell and tube condenser," where the steam comes in contact with banks of tubes that have cold water going through them, and the steam condenses to water. In both cases the water is pumped out of the condenser to a cooling tower.

The change of design was due to the need for hydrogen sulfide abatement. There is a small amount of hydrogen sulfide in the steam, and that creates a potential odor nuisance problem, and there is an ambient air quality standard for hydrogen sulfide emissions that must be met to avoid annoying our neighbors. The initial plants had no abatement systems, and to be allowed to develop the Geysers field, they had to add

(continues)

(continued)

abatement systems to the existing plants by retrofitting them with abatement systems. The new plants had to be designed to have abatement from Day 1. It turns out that the shell and tube condenser allows you to extract the gases in a more concentrated form and pipe them to an abatement system more readily than if all the cooling water mixes with the steam, which makes it more difficult to abate the H_2S.

Q: Are the new plants also more efficient thermally?

A: Yes, there have also been some improvements in low-pressure steam turbine design over the years, so the newer plants have more efficient turbines. We've also retrofitted some of the oldest plants with state-of-the-art steam turbines to extract the maximum amount of energy out of the steam.

Q: The plants are not all of the same size. What accounts for the differences in size as time went on and more plants were built?

A: The very first two plants were 12.5 MW and 14 MW capacity, and the utility that built the plants wasn't too sure about geothermal energy so they hedged their bets by making them smaller. As the steam-field developers demonstrated ongoing success in drilling high flow-rate steam wells, the utility was willing to invest in larger power plants designed to operate for 30 years. There are a few localized areas at the Geysers that have thermal features, evidence of surface leakage of steam from the reservoir below—hot springs, fumaroles—and that's where the first wells were drilled, adjacent to those thermal areas. But they started stepping out in quarter-mile increments to see if the reservoir down below was bigger than that. They kept stepping out and stepping out on this trial-and-error basis and kept hitting steam until they had—over several decades—defined an area that turned out to be about 40 square miles. Wow! But the majority of the field here has no surface manifestations of hot springs, fumaroles, etc., that would give any indication that there is this hot reservoir below that is full of steam.

Q: I've noticed that the power output of the Geysers has fluctuated over the years. First, of course, rising quite rapidly but then tailing off. What accounts for that?

A: There was a boom time for renewable energy back in the '70s in response to the OPEC "oil crisis" and we had many companies all competing to rapidly develop the Geysers. As a result too many "straws" were being put into the same steam resource, and the output from the Geysers peaked in about 1989. Over time, they've found that they had overdeveloped the field for how much fluid is in place and for how long those flow rates can be sustained. So there was some drop off in steam production and output. Some of the plants needed to be retired. That hints at one of the challenges of geothermal energy: There is a reservoir engineering aspect of being able to determine how big the geothermal resource is, how much fluid is in place, and how long it can last. Over these last three decades there have been advances in geothermal reservoir engineering that have given engineers a better way of determining how much is there and how long it will last. But those engineering skills were in their infancy back in the '70s, when all of this development happened. They didn't really have a good way of determining how big the steam reservoir was at depth, and so that is what led to the overdevelopment. There used to be a rough rule of thumb of about 40 acres per well, but that might be too dense or it might not be dense enough. It depends on how deep the resource is and so on.

Q: Over the years have you managed to bring the cost per kilowatt-hour down? Have there been changes in terms of the economic efficiencies at which the plants are operated?

A: Early on, there were companies that operated the steam field, and there were other companies that operated the power plants, and those two might have had incentives to work at cross-purposes with one another. In 1999, during deregulation of the electricity industry in California, Pacific Gas and Electric sold its geothermal power plants to Calpine, and

(continues)

(continued)

Unocal sold its steam fields to Calpine. So we now have the steam fields and the power plants under one management, and we are now able to optimize the resource and match it to the power plant and improve our efficiency. That also allows us to reduce our management and manpower costs. We've reduced the costs by consolidating under one management and reducing the manpower needed to run this whole facility.

Q: In a conventional fossil fuel plant, there would be people concerned with providing the fuel, moving the fuel, managing the boiler, and all of that—

A: Yes, for a natural gas–fired power plant—you have the gas fields, the gas wells, and gas pipelines, which are very far away, and you don't see those. They arrive underground at the power plant. There might be a natural gas–fired power plant on 11 acres, let's say, and a fairly small crew that operates that power plant. Here, we have all the wells, pipelines, and all the facilities all in one place. So you see everything related to the geothermal production all in the same area. For a natural gas–fired plant all those facilities that provide the fuel are hidden and far away. The overall footprint of a geothermal facility is actually smaller.

Q: Sure. The supporting infrastructure (for a natural gas– fired power plant) is distributed over the whole globe sometimes. With respect to a geothermal power plant, how many engineers would it take to operate a plant?

A: It varies with the geothermal resource and over time as the facility goes through its life cycle. We need engineers for all phases of the operation during the life of the facility. All of the equipment must be designed, constructed, operated, and maintained through the life of the equipment. Engineers play different roles during all of those phases. You have reservoir engineers, drilling engineers, and geologists focused on resource characterization, modeling, and forecasting. On the operations and facilities side, you have chemical, mechanical, and civil engineers, and maybe some electrical engineers dealing with all those kinds of issues. And then to operate and maintain the facilities we need trained operators, mechanics, and technicians to run the facility and keep it operating efficiently throughout its life.

Q: With respect to the future of the Geysers, when power production started dropping off because it was overdeveloped, the operators at the time—and I don't know if that was Calpine—started to inject treated wastewater from nearby municipalities. Has this stabilized the output of the Geysers?

A: Yes, injection of additional water has helped to stabilize the output of the Geysers. We started injecting water from Lake County in 1997, at about 8 million gallons per day. Then, at the end of 2003, we started injecting an additional 11 million gallons per day from the Santa Rosa area. So a combined 19 million gallons per day (about 13,000 gallons per minute) is distributed across the Geysers field. We've successfully stabilized the output at about 6 million megawatt hours of electrical generation per year for the last six years. This is enough electricity to supply the annual needs of over 1 million people.

Q: How long do you expect to operate the Geysers at this rate? Can you maintain this permanently?

A: For the long-term there will be some ongoing decline. We look at capital investment projects, adding wells, modifying equipment in plants, and trying to maximize how much we get out of the steam to continue to keep the output at the current level. We're doing a pretty good job of it currently, and we have teams of engineers and geologists working to figure out which projects would help us maintain the current output.

Q: I'm not clear about what the challenges are. Is it that water is not being distributed through the system? Is the formation itself cooling off? What are the challenges you face as you work to maintain production?

A: The key challenge is managing the steam resource. Water can short-circuit and go from an injection well right over to a steam well. That's called "breakthrough," and that can stop steam production in adjacent

(continues)

(continued)

wells. You have to be careful to avoid that. We can inject too much in an area and that can, as you mentioned, "cool" an area. So we monitor the output in adjacent steam wells to see if the steam production is going up or stabilizing or is actually dropping off. By careful management over where we put water and how much—through watching the output of the nearby wells—we can optimize where we put it and the results. In some cases we were putting too much in one area and we realized that we were hurting the steam production and the plant output dropped off. Then we back off on injection into that area, and let steam production recover for awhile. We have about 350 steam wells and about 60 injection wells, and we have a computer system that monitors flow rates, temperatures, pressures, and the plant performance, and all the equipment in the power plant, and we are able to track the performance of these different plants and wells and try to optimize it.

Q: That's fascinating. Calpine manages all the plants as a single network?

A: Yes, we operate 17 power plants at the Geysers and they are interconnected with crossover pipeline so if one shuts down we are able to transfer the steam to an adjacent power plant. That adds to our reliability and how much we are able to produce, but it also helps us minimize steam emissions. It prevents us from wasting the steam, and also prevents us from emitting some of those gases that have to be abated.

Q: What would be the single largest challenge in maintaining output from the Geysers?

A: The biggest challenge is managing that water. The biggest single cost is drilling a well. Each steam well at today's prices is over $4 million. We have about 350 steam wells and 60 injectors. So we need to make sure that we protect those wells, monitor how they are doing, and do repairs on some of them. We could not afford to drill all of those wells at today's prices.

Q: This has been fascinating. I very much appreciate your time and insights on this.

(continued from page 131)
the working fluid absorbs the heat from the water and changes from a liquid to a gaseous phase.

The choice of working fluid is crucial to the success of the design. The amount of heat that flows from the hot water through the heat exchanger into the (liquid) working fluid is fixed by the temperature of the water and the rate of flow of each fluid. This energy must be sufficient to cause the working fluid to change phase from liquid to a vapor. It is the expanding vapor of the working fluid that drives the turbine-generator unit. Once past the turbine, the vapor passes through a condenser, where it again becomes a liquid, and the liquid is pumped back to the heat exchanger, where the cycle repeats. Meanwhile, the water continuously circulates out of the thermal reservoir, through the heat exchanger, and back down into the thermal reservoir.

The maximum thermal efficiency attainable by the binary system is the lowest of the three systems, typically about 10 percent. As a consequence, more water, when measured by mass, must be circulated in order to produce the same amount of electricity. The need to maintain a higher flow rate to obtain the same power output and the additional complexity of the plant drive up costs. This technology is employed, for example, at the Casa Diablo geothermal field in California, at facilities owned by Mammoth Pacific, LP.

COPRODUCTION OF GEOTHERMAL ENERGY

Today, as energy prices soar, many nations, the United States included, seek new ways of rapidly producing large amounts of inexpensive power domestically, and geothermal energy is being reevaluated. The big questions about whether thermal energy can be mined from formations of hot, impermeable dry rock located deep within Earth's interior at a price that is competitive with other alternatives remains open. This is the main question because most geothermal energy lies within this type of formation. There is, however, a smaller question of whether it is possible to quickly expand geothermal energy production in some way using only the technology at hand, and to that question the answer seems to be yes.

In the United States, the key to a rapid expansion of geothermal energy production may lie with the oil and gas industry. Within the contiguous 48 states, all of the easily exploited deposits of oil and gas have been located and, for the most part, exhausted. As a consequence producers are drilling ever deeper into Earth's interior in search of oil and natural gas, and as they produce natural gas and oil they are also producing large amounts of water. (Water, oil, and natural gas are often found together, and in order to obtain one, it is often necessary to extract them all.) The amount of water produced along with the petroleum is staggering. In the lower 48 states, more than 40 billion barrels of water are produced from oil and gas wells each year. This water, often polluted by oil, currently represents a disposal problem for the producers.

The production of geothermal power requires large amounts of fairly hot water. To date, there has not been sufficient research to characterize the temperature of the water produced from oil and gas wells, but it is certainly known that at least some wells are producing water at temperatures that could be used by a binary geothermal system. If electricity were produced in conjunction with the production of oil and natural gas, it would be classified as a coproduct of oil and gas production.

The attraction of the idea rests on two observations: First, much of the infrastructure required for the production of geothermal energy is already in place at an oil and gas well. The well is in place, of course, but the permeability of the reservoir rock has also already been addressed. The rock may have been permeable at the outset, but if the reservoir rock from which the oil, natural gas, and water is being withdrawn was initially impermeable, it has already been hydraulically fractured to facilitate the flow of oil and gas (otherwise the oil and gas would not be flowing). The production of electricity makes the operation of the well more profitable. The authors of the influential 2006 study, *The Future of Geothermal Energy: Impact of Enhanced Geothermal Systems (EGS) on the United States in the 21st Century* described the advantages of the coproduction of electricity in the following way: ". . . piggybacking on existing infrastructure should eliminate most of the need for expensive drilling and hydro-fracturing operations thereby reducing the risk and the majority of the upfront cost of geothermal electrical power production."

Second, any hot water produced at the well is currently an economic loss, because it is solely a waste disposal problem. But with a geothermal power plant, it can, at least in theory, be turned into a source of profit. The data needed to evaluate the economic feasibility of coproducing geothermal energy together with oil and gas production is relatively easy to acquire. Efforts are already under way. This may be the first major new and profitable application of geothermal energy in decades.

Two Other Geothermal Technologies

Energy exists in many forms. The two forms of energy of most interest here are electrical and thermal, and of these two forms of energy, electrical energy can be used in more ways and in more places. Electricity can be transmitted across large distances with little loss. It can (and is) used to heat homes and businesses and to power most of the appliances that make modern life possible. For these reasons, geothermal energy is converted into electrical energy whenever it makes economic sense to do so.

Whether or not it makes sense to convert geothermal energy into electricity depends on many factors, but one of the most important is the temperature of the geological formation of interest. As mentioned in chapter 10, heat engines are designed to use the temperature drop between the environment and the thermal reservoir, and the smaller the temperature drop, the less efficient the engine (all other things being equal). As the temperature drop

Mammoth Hot Springs, Yellowstone National Park. Some geothermal resources are easy to identify. Most are not. (*J. K. Hillers, USGS*)

approaches zero so does the percentage of thermal energy that can be converted into electricity. This is not to say that it is impossible to build a heat engine that can utilize a small temperature drop. A few such engines have, in fact, already been built and tested, but from a practical point of view, the bigger the temperature drop the better.

While the exact temperature at which a geothermal resource ceases to be economic for purposes of electricity generation depends somewhat on the technology used to exploit it, there are no geothermal power plants that produce electricity from a thermal reservoir with a temperature less than 212°F (100°C), the temperature at which water boils at atmospheric pressure. But there are many important nonelectrical applications that make use of the thermal energy in water even when the temperature of the water is less than the boiling point: Home heating, washing, and other such applications often depend on water that is hot but not boiling. This chapter describes two geothermal technologies that exploit more modest temperature differences between the heat source and the environment. In particular, they do not involve the conversion of thermal energy into electrical energy.

The first application, direct-piped hot water, makes use of warm water pumped from the ground and used near the location of the well, often for heating. The second application uses a device called a *geothermal heat pump,* a piece of equipment that could drastically reduce the energy requirements of many future homes and business should it become widely adopted. The value of these technologies lies in the fact that they exploit low-grade thermal reservoirs, which are often easier to access and which are distributed more widely than are thermal reservoirs capable of supporting electricity production.

DIRECT-PIPED HOT WATER

The simplest way to use water that is warm, but not warm enough for generating electricity, is space heating. The idea is simple enough: Drill a well into a supply of warm water, pump the water out of the ground, circulate the water through a radiator or other type of heat exchanger, being careful to extract as much heat as possible, and dispose of the cooled water. The value of the resource depends on the initial temperature of the water, the distance between the well where it is produced and the location where it is used, and the amount of hot water that can be sustainably extracted. This is an example of so-called direct-use technology.

The simple direct-use scheme just described can be improved in several ways. First, it is important to keep in mind that geothermal energy projects are not renewable in the sense that solar energy is renewable. One cannot exhaust the supply of energy from the Sun, but it is possible to exhaust the supply of water from a geothermal energy source by removing it faster than it can be replenished.

Consequently, where it is possible, designers attempt to protect the supply of water used to transport the thermal energy to the surface. One method is to use the water, and then inject it back into the formation from which it was taken. The injected water circulates through the porous hot rocks where it is reheated and made ready for reuse. The other method, popular where the supply of heat is very near the surface, is to place a heat exchanger in the ground and circulate water through the heat exchanger. The water absorbs sufficient thermal energy to make it valuable, at which time it is ready to be used. These techniques can extend the life of the resource.

A second way to improve direct-use technology is to carefully insulate the pipes used to transport the hot water from the wellhead to the point of use. There is currently no way of transporting thermal energy with the ease that electrical energy is transmitted, but by carefully insulating pipes it is possible to transport the working fluid, which is almost always water, tens of miles without significant losses, provided large-diameter, well-insulated pipes are used, and the rate of flow is kept high. To see how this works in practice, consider two of the most advanced systems for using low-grade thermal energy, the first of which is located in Iceland and the second of which is located in France.

Iceland is justly famous for its use of geothermal energy, and direct-use applications are common. In fact, direct-piped heating is the main application for geothermal energy in Iceland. (Most of Iceland's electricity needs are supplied through hydroelectric power.) Almost 90 percent of all buildings are heated with geothermal energy, and its widespread use reflects the peculiarities of Iceland's geography. The population of Iceland is about 313,000 people, and they inhabit an island of about 36,700 square miles (103,000 km²), an area about the size of the U.S. state of Maine. The island is located on the Mid-Atlantic Ridge, a zone where two adjoining tectonic plates are moving away from each other. As a consequence, Earth's crust is especially thin and vast deposits of magma lie relatively close to the surface. Direct-piped hot water heats virtually all of the buildings in Iceland's largest city, Reykjavík, where about half of the nation's people live, and direct-piped water is used in many isolated rural homes as well. The technology is relatively simple and afford-able because so much geothermal energy is so close to the surface. In more urban centers, direct-piped hot water is even used to melt the snow on sidewalks and streets and to heat outdoor swimming pools, which are common in this cold land. A great deal of produce is grown in Iceland in greenhouses that are heated with direct-piped water, and other industries also make use of this abundant and inexpensive resource.

But while it is easy to celebrate such inventiveness, it is difficult to draw broad conclusions from Iceland's experience because of the special circumstances that prevail there: a relatively small popula-tion, concentrated in a few urban centers, situated on one of the greatest sources of geothermal energy in the world.

France's experience with geothermal energy is, perhaps, more instructive. There are two areas in France where low-grade ther-mal resources are available, and both are exploited via direct-piped hot-water technology. Most efforts are concentrated in the Parisian Basin, a large geographical feature that includes Paris, and the geothermal energy obtained from the Basin is used for residential heating. A smaller program exists in the Aquitaine Basin, a geologi-cal feature located in southwestern France. The Aquitaine projects are more diversified and include, in addition to residential heating, the provision of direct-piped heat for agriculture and fish farming projects. Although France tested geothermal energy projects prior to the 1973 oil crisis, after the crisis efforts to use geothermal energy were accelerated. Today, low-grade thermal reservoirs are used to heat approximately 200,000 French homes with water that has an initial temperature of 140°F–180°F (60°C–80°C).

Finally, there are a number of cities in the United States that make use of low-grade thermal reservoirs. Idaho, for example, has abundant geothermal resources. A number of greenhouse-type operations in that state are heated with direct-piped water, for example, and aquaculture operations use direct-piped geothermal energy to grow species that are not native to the area. (Alligators are raised in Idaho on farms that depend upon direct-piped hot water.) In Boise, site of the nation's first geothermal heating district, a few hundred homes, some government buildings, and a few businesses currently use direct-piped systems. (As mentioned in chapter 8, direct-piped heating was begun in Boise in the late 19th century.) Other applications of direct-use thermal energy in the United States are relatively small in scale, in part, because the initial costs of geothermal energy, even "simple" direct-piped systems, tend to be high. Once installed, however, the costs of these systems are reasonably predictable, and properly maintained, they provide a reliable and long-lived source of hot water.

GEOTHERMAL HEAT PUMPS

Geothermal heat pumps use the thermal properties of the ground in a way that is very different from those of geothermal generating stations or direct-use technologies. The problem with the direct-piped hot-water applications and with geothermal power stations is that both depend on ready access to concentrated deposits of thermal energy. By contrast, given enough land, geothermal heat pumps, which can heat and cool buildings, can be successfully installed almost anywhere.

To understand how a heat pump works, keep in mind that heat always flows spontaneously from warmer to cooler regions. A heat pump is designed to reverse the process: It "pumps" thermal energy from a cooler region to a warmer region—that is, it reverses the natural flow of heat. The ideas involved are not new. Refrigerators, for example, move thermal energy from a cooler region (the freezer) to a warmer region (the air in the kitchen) using a process that requires the expenditure of considerable amounts of energy.

Part of a geothermal heat pump. Heat pumps are often cost effective to install in a new house, but they are usually very expensive to retrofit. *(Integrated Design Associates, Inc.)*

When heating a home, a geothermal heat pump performs the same function as a refrigerator, transferring heat from the cooler ground to the warmer air inside the house.

The key observation on which geothermal heat pump design rests is that several feet below the surface of the ground the temperature of the soil remains constant throughout the year. The depth at which the temperature first ceases to vary depends somewhat on the location of interest, but the variations are not large. In temperate regions, the temperature of the ground at a depth of about 10 feet

(3 m) is a constant 40°F–50°F (5°C–10°C). A residential heat pump operates between this region of constant temperature and the air inside a home or business.

A heat pump is usually designed to operate in two "directions." During the winter months, it extracts thermal energy from the ground and transfers it to the air inside the building—that is, it takes the place of a furnace. During the summer months, a heat pump can extract thermal energy from the air inside the building and transfer it to the cooler ground—it takes the place of an air conditioner. The heating and cooling functions require energy— heat pumps consume rather than produce energy—but it usually takes substantially less energy to operate a heat pump than it does a furnace or an air conditioner. Because heat pumps consume significantly less energy than the alternatives, they are advantageous to use because they free energy for other applications—energy that would have been used for heating and cooling.

The component that is unique to a geothermal heat pump is a large loop of piping that is buried beneath the ground. The geometric configuration of the loop depends on local conditions, but the system performs the same basic way regardless of its geometry. Fluid, either water or an antifreeze solution, circulates through the piping. When the heat pump is used to heat the house—and to keep the discussion specific this is the only case that will be considered for now—cool fluid is circulated through the loop. (This is fluid that is cool *relative to the temperature of the ground*.) As the fluid passes through the loop, thermal energy flows from the warmer ground to the cooler fluid because thermal energy spontaneously flows from warmer to cooler regions. By the time the fluid has circulated throughout the loop it is at roughly the same temperature as the soil. This warmed fluid is pumped inside the building, where a refrigerator-like piece of machinery extracts thermal energy from the fluid and transfers it to the air in the building. Extracting thermal energy from the fluid in the loop causes the temperature of the fluid to drop. Finally, the cooled fluid is pumped back through the loop in order to absorb more thermal energy from the ground.

The system can be operated continuously for long periods of time because a good deal of thermal energy can be removed from the ground before there is a substantial change in ground temperature. Engineers and scientists describe this situation by saying that the ground has a high specific heat. (See chapter 5 for a discussion of specific heat.) As thermal energy flows from the ground to the fluid in the loop, there is a slight drop in ground temperature near the loop, causing thermal energy in regions farther away to flow toward the loop, thereby replacing the energy lost to the fluid in the loop. The reason for the flow of heat (again) is that heat spontaneously flows from hot to cold. Even though the system may be operated continuously for prolonged periods, the temperature of the ground will remain virtually unchanged in such a system.

Heat pumps also operate between the air inside a building and the air outside, and they are designed according to the same general principles as geothermal heat pumps. The difference between the two types is that the temperature of the air outside a building can vary greatly from season to season. When the temperature difference between the air inside the house and the air outside is very large, forcing thermal energy to flow from cold to hot requires a good deal of energy. This is why geothermal heat pumps are superior: The temperature of the ground is both steady and moderate, and as a consequence, a geothermal heat pump requires much less energy to pump adequate supplies of heat inside the house than a heat pump operating between cold outside air, for example, and inside air that is already at a moderate temperature.

During the summer months, most geothermal heat pumps can extract thermal energy from the air inside a building and deposit the energy in the ground outside via the same loop of circulating fluid. Conceptually, this amounts to running the heat pump in re-

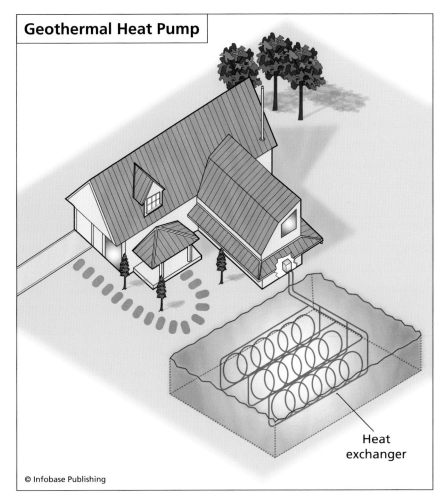

Geothermal Heat Pump

Heat exchanger

© Infobase Publishing

A geothermal heat pump system is a very energy-efficient way to heat and cool. *(Source: EERE)*

verse, although the actual procedure is accomplished in a somewhat different way.

The United States has long accounted for most of the geothermal heat pump market, but they have become increasingly popular in many areas of the world. Their value lies in the fact that they dramatically reduce the amount of energy required for heating and cooling, thereby freeing energy for other uses. As a nation's economy grows, certain sectors of its economy will require additional energy. There are two ways to meet this demand. The most obvious way is to build more power plants. Geothermal, hydroelectric, coal, nuclear, natural gas, and so on . . . there are many ways to *generate* additional energy, but each generating station represents an additional expenditure and an additional environmental cost. The other way to meet increased demand in one sector of the economy is to use less energy in another sector and transfer the difference. Geothermal heat pumps make such transfers possible because they perform the same function as a furnace or air conditioner while using between 30 percent and 70 percent less energy. As a consequence, they free substantial amounts of energy for use elsewhere. In this way conservation efforts contribute to the energy supply.

The most serious disadvantage of geothermal heat pumps is that they tend to be more expensive to install than more conventional heating and air conditioning units. As energy prices continue to rise, however, that difference will become less significant. This is a technology that will almost certainly become more widespread in the future.

The Economics and Environmental Impacts of Electricity from Geothermal Sources

Before any energy source becomes an important part of the energy market, it must be produced at a cost that consumers are willing to pay. Sophisticated science and engineering are seldom sufficient to guarantee that a particular technology will be adopted. While some science is pursued independent of its "practical" value, this is rarely the case with respect to energy, where technologies are judged not just on their own terms but also in comparison with other competing technologies. With respect to energy production, the stakes, economically and environmentally, are too high for it to be any other way. Dr. Stan Bull, the former director of research and development programs at the U.S. National Renewable Energy Laboratory, one of the nation's premier government laboratories, was

Collecting gas samples at Momotombo geothermal field *(Cosmogenic Isotope Laboratory, Rochester University)*

very clear when describing the nature of the work conducted at his facility: "Everything we do is intended for industry." This chapter examines some of the economics of geothermal power generation.

In addition to science, engineering, and economics, environmental concerns have come to play a role in determining whether a particular technology will succeed in the marketplace. To be sure, the importance policymakers claim to assign to environmental concerns is often overstated. Severe environmental, health, and safety issues associated with the production and consumption of coal and oil are well-documented, for example, but most countries have, at least so far, made only modest efforts to reduce their consumption of these fuels. The same statement applies to individuals. Even those who claim to oppose the consumption of fossil fuels continue to drive cars, ride planes, and use electricity produced at coal-fired power plants. Faced with a choice between energy and the environment, most people choose energy most of the time. Nevertheless, the importance assigned to environmental questions has increased as the consequences of using today's power-generation technologies have become more apparent. This chapter examines some of the environmental implications of geothermal energy.

ECONOMIC COSTS AND THE PROBLEM OF SCALE

Geothermal power plants are not simply technical achievements. They are built to earn investors a profit. Geothermal power plants are operated in a way that maximizes investors' returns, and the profit motive also helps to explain why there are not more geothermal plants.

A geothermal power plant is only expensive or inexpensive relative to some other power-generation technology. To make these types of comparisons, it helps to know a little about how utilities incorporate different technologies to meet the demand for electricity. More than a century of experience has demonstrated that demand for electricity fluctuates according to the season, the day of the week, the time of the day, and a number of other well-known factors. Some fluctuations are predictable. Demand tends to be higher during the middle of the day than during the middle of the night. Other fluctuations in demand are more random. System operators know, for example, that on very hot days, more electricity will be needed than on cooler days because air conditioning is a very energy-intensive technology. What they do not know far in advance are the dates on which hot days will occur.

For each particular electricity market, there is also a certain minimum amount of power that needs to be produced regardless of the weather or the seasons or any other random or cyclical factors. Hospitals, certain manufacturing concerns, emergency services, and traffic lights, for example, all consume power 24 hours per day at rates that are highly predictable. These demands form a certain base power requirement. While there will be times when each utility is called upon to produce power in excess of this minimum demand, it will always have to produce enough power to meet the minimum demand, 24 hours a day, seven days a week.

Minimum demand is called base load power and fluctuations above the minimum are called peak load power. (Sometimes a third category, intermediate power, is used to describe certain intermediate fluctuations, but in this volume, as in many other descriptions, all fluctuations above the base load will be called the peak load requirements.)

Coal and nuclear power plants are most often used to meet base load power requirements. The reasons are technical and economic. Coal and nuclear plants are inexpensive to operate, they are highly reliable, and they work most efficiently when operated at steady power levels for prolonged periods of time. They are, in short, best suited to meet base load requirements. Most peak load requirements are met with natural gas plants, because they are easy to start and easy to shut down. Natural gas plants could also be used to meet base load requirements, and they have been used that way in the past. Some are still used that way. But natural gas has become increasingly expensive, and using a natural gas plant to provide base load power results in high electricity bills. Hydroelectric power is another technology that can be used to meet either base load power or peak load power requirements. The way a particular hydroelectric facility is used often depends on a complex mix of economic and environmental concerns. In the United States and most other nations with large economies, almost all of the demand for electricity is met with some combination of these four technologies.

Geothermal power is ideally suited to provide base load power. Its energy source, the enormous volumes of hot rock buried deep within the planet, is, when properly used, almost inexhaustible, and the technology used to convert thermal energy to electrical energy is very reliable. Except for the fuel source, the technology used in geothermal plants is essentially the same technology as that used in coal, natural gas, and nuclear plants. Engineers use a measure called the capacity factor of a plant to describe the way that a plant is operated in practice. The capacity factor is defined as the quotient of the actual power output of the plant over a representative period of time (often one year) to what the output would be if it were operated continuously at full power over the same period:

(capacity factor) = (actual output) ÷ (theoretical full power output over same period of time)

Because the actual output cannot be greater than the full power output over the same period of time, the capacity factor is a number that lies between zero and one. In the United States, coal-fired plants generally have capacity factors of approximately 70 percent, and nuclear plant capacity factors hover around 90 percent. Geothermal power plants typically have capacity factors in excess of 90 percent. They generally shut down only for routine maintenance. (By contrast, photovoltaic systems generally have capacity factors of between 20 and 30 percent.)

Geothermal power projects usually consist of multiple, relatively small generating stations distributed over a large geographic area. Each station is situated over a different part of the resource—that is, the formation of hot, permeable, water-saturated rock—in order to use the thermal energy at the site below it. No single geothermal power plant can operate at a scale similar to that of a large coal-fired or nuclear plant without damaging the resource on which it depends. Producing large amounts of power from a single enormous plant would entail withdrawing large amounts of water or steam quickly from too small a volume of rock. Either there would be insufficient water or steam to continue the practice for long, or the practice would create a zone of cool rock in the much larger hot rock formation. In either case, the power plant would soon be unable to produce power at its rated output.

Because geothermal facilities are comprised of numerous smaller generating stations—where "small" is relative to many coal and nuclear facilities—they do not benefit from economies of scale. For these reasons, and because drilling costs are always high, geothermal power plants tend to be fairly expensive to build.

The cost of constructing a power plant is sometimes characterized by a quotient: the total cost of "overnight" construction (a term that means that financing costs are not taken into consideration), divided by the plant's rated power measured in kilowatts.

cost = (overnight construction costs) ÷ (rated output in kilowatts)

So, for example, a power plant that had an overnight construction cost of $100 million and a rated output of 100 MW—which equals 100,000 kW—would have a cost of $1,000 per kW ($100,000,000 ÷ $100,000kW = $1,000 / kW$).

By measuring the cost of a plant in this way, one can compare the value of an investment in one project versus another without reference to any particular technology. Geothermal power plants have an overnight cost of about $2,700 per kilowatt. This is a reasonably accurate measure since some plants are currently under construction, which means that figures can be gleaned from current practices. But $2,700 per kilowatt is four to six times as expensive as the cost of a combined-cycle natural gas plant, which is currently the most efficient type of gas plant available. (Comparing geothermal power plants with other base load power producers, namely nuclear and coal-fired power plants, is more difficult because no new nuclear power plants have been built in the United States in decades and the nation's coal-fired power plant fleet is also very old. Estimates of the cost of deploying modern coal-fired and nuclear plants depend on various difficult-to-test assumptions and vary substantially.)

Historically, power producers have been reluctant to develop geothermal power because of the high costs involved. These costs are partially offset by the fact that the heat source, geothermal energy, is free. But accessing the heat source and maintaining the supply of hot water is not free—and is, in fact, fairly expensive. Drilling is always expensive, and wells, once drilled, need to be maintained, and water, once withdrawn, must be injected back into the ground to maintain the system. Even at today's high energy prices, geothermal energy cannot be characterized as a bargain.

Costs can, of course, be defrayed by subsidies, which are always offered at the expense of taxpayers. These subsidies, when applied to geothermal power plant construction, are not currently a source of concern because of the modest size of the geothermal power sector, but if the size of that sector were to substantially increase, which it must do if geothermal energy is to make a larger contribution to the energy markets, so would the subsidies. At that point taxpayers would become more involved in the discussion of whether geothermal energy is the best use of taxpayer funds. (Keep in mind that in order for the geothermal sector to make a larger contribution to the total energy supply, it must grow faster than competing technologies. It is not enough that the sector simply grows in size, since the market of which it is a part is also growing.)

There is one additional barrier to the development of geothermal power that has not been mentioned. Because geothermal technology currently focuses on the development of formations of hot rock that are permeable and water-saturated, power producers have a fairly short list of sites that can be developed, because locations where all three factors come together are unusual. Only some of these sites are close to high-voltage transmission lines. Without access to high-voltage transmission lines, a modern power-generating station is useless. Unfortunately for proponents of geothermal power, the construction of new high-voltage transmission corridors has become increasingly difficult. Opponents include private landowners, who object to selling their land, and individuals living near sites who do not want to see high-voltage lines near their homes.

(continued on page 164)

Global Warming and Geothermal Power

To see the effect of Earth's atmosphere on Earth's surface temperature, consider the Moon. The distance from the Sun to the Moon is very nearly equal to that of the Earth-Sun distance and so over the course of a year both bodies receive roughly equal amounts of solar energy per unit area, and yet some places on the Moon are much hotter, and some are much colder, than anywhere on Earth. For a long time the explanation for the difference was both easy and unexamined: Earth has an atmosphere, and the Moon does not. Earth's atmosphere filters out some of the Sun's energy before it reaches the surface, and it traps some of the energy that does reach the surface, thereby moderating the temperature changes associated with the transition from day to night. By contrast, at any given time, part of the Moon receives the full brunt of the Sun's rays, and at night the thermal energy absorbed by the Moon's surface radiates into space unimpeded. As a result lunar daytime temperatures are as high as 250°F (123°C) and nighttime temperatures as low as –450°F (–233°C). And while these very broad differences between the surface temperatures of the two bodies have been known for a long time, scientists were, until recently, less attuned to the differences small changes in the chemistry of Earth's atmosphere could make to the rate at which heat is retained.

Earth's atmosphere is, when measured by mass, almost 76 percent nitrogen. Almost all of the rest—approximately 23 percent—is oxygen. (All figures in this section are for the "dry" atmosphere—that is, they do not take into account the contribution of water vapor.) About 1 percent of the dry atmosphere is argon. Other gases make up such a small percentage of the atmosphere that when all figures are rounded to two significant figures, as was done here, their sum is less than the round-off error. Carbon dioxide and methane are two of these minor gases.

What makes carbon dioxide and methane important is that even when they are present in small quantities, they cause the atmosphere to retain significant amounts of thermal energy. If the percentages of carbon dioxide and methane in the atmosphere were substantially reduced, much of the thermal energy that is presently retained by Earth's atmosphere would radiate back out into space, and the planet would be a much colder place.

Modern societies, which are highly dependent on fossil fuels, emit large amounts of carbon dioxide directly into the atmosphere as a by-product of burning these fuels. Much smaller amounts of methane are emitted as a result of human activity, but because methane is about 20 times as efficient as carbon dioxide at retaining heat, even these small amounts of methane affect the thermal properties of the atmosphere. (Water vapor, which is not specifically addressed here, also retains significant amounts of thermal energy that would otherwise radiate into space.)

Methane does not last very long in the atmosphere, but carbon dioxide is more persistent, and such large amounts of it are emitted that the oceans, which absorb a great deal of carbon dioxide from the atmosphere, may become saturated with it. There is even some evidence, so far inconclusive, that this has begun to occur. If this proves true, then even if carbon dioxide emissions stabilized, the concentration of carbon dioxide in the atmosphere could be expected to continue to increase, perhaps even at a faster rate, since less would be absorbed by the oceans. And even if the oceans' ability to absorb carbon dioxide has not been compromised, modern societies are so heavily dependent on fossil fuels that atmospheric levels of carbon dioxide can be expected to continue to increase well into the future since carbon dioxide is currently emitted faster than the oceans (and green plants) can absorb it. And it is widely acknowledged that the rate at which carbon dioxide is emitted is expected to increase over at least the next decade as more people adopt more energy intensive lifestyles. Climate change, a consequence of the slow but steady increase in the concentration of carbon dioxide in the atmosphere, can be expected to continue and even accelerate into the indefinite future.

Because the heat-retention properties of the atmosphere are changing, many nations are engaged in research into finding alternatives to fossil fuels. This is one reason why geothermal energy, which produces few to no emissions and is, in theory, a good alternative to coal and natural gas, has begun to attract the attention of many nations.

(continued from page 161)
Others object to the construction of such corridors based on the impact that the high-voltage power lines or even the corridor itself may have on scenery, parks, or wildlife. No power producer would build a plant without guaranteed access to the necessary high-voltage transmission lines. Obtaining that access where none currently exists can take years. (See the sidebar "Connecting Producers and Consumers" in chapter 6.)

Despite its environmental and technical advantages—and there are several and they are substantial—geothermal power can be expected to develop slowly in the United States and in many other areas of the world for the next several years at least.

GEOTHERMAL ENERGY AND THE ENVIRONMENT

Geothermal heat and power technologies are local in nature. The type of plant chosen for use, the rated power of each plant, and the way that water resources are managed are largely determined by local conditions. Because each facility is unique, it is difficult to make very specific statements that are true for all systems, but some general statements can be made about the way that the operation of this class of power plant affects the environment.

The emissions produced by geothermal power plants depend on the characteristics of the steam produced at the well and the design of the plant. As water circulates through a geothermal reservoir, it can absorb various materials, some of which can pose environmental hazards when the water is finally brought to the surface. In particular, if steam from geothermal sites is released into the atmosphere, as is done at some geothermal plants, significant amounts of hydrogen sulfide and carbon dioxide could be released together with the steam. These gases are frequently dissolved within the water produced at the well. In fact, some early geothermal power plants produced significant quantities of sulfur dioxide during the normal

The cooling tower for the West Ford Flat power plant, which is part of the Geysers, the world's largest geothermal power development *(The Geysers)*

course of their operation. Today, the technology for eliminating hydrogen sulfide emissions is widely available. At the Geysers, for example, about 15 short tons (14 metric tons) of sulfur are captured each year for each megawatt of electricity produced, and emissions of sulfur dioxide from the Geysers are now almost zero. With respect to hydrogen sulfide, emissions from geothermal power plants are always miniscule compared with those produced by similar-sized fossil fuel plants regardless of the fossil fuel considered, but as with fossil fuel plants, carbon dioxide is generally emitted during the course of power production.

In a binary system, steam is not emitted to the atmosphere. The hot water is, as previously described, kept under the operator's control at all times. It is produced at the well, piped to a heat exchanger, and injected back into the ground to form a closed loop. Binary power plants, therefore, produce power with essentially zero emissions. As a general rule, no matter the type of geothermal power plant considered, geothermal energy is remarkably clean when measured by the emissions produced.

As mentioned earlier, geothermal power plants are not renewable sources of energy—at least not in the same way that solar resources are renewable—and so it is instructive to compare the environmental impacts of geothermal plants with solar power plants. Solar energy is highly dispersed. There is, to be sure, a great deal of energy in the Sun's rays, but there is only a modest amount of energy at any given location. Because solar energy is dispersed over large areas, exploiting it on a commercial scale requires very large areas of land. Producing commercially significant amounts of electricity from solar power will, therefore, require numerous and very large fields of solar collectors. There is no other way to obtain this energy. By contrast, geothermal power plants occupy comparatively small plots of land. They require large volumes of hot rock to provide the hot water on which they depend, but this rock is situated far underground. The power plants simply sit at the top of multiple very long pipes, some of which extract and some of which inject water into the hot rock below. Land use requirements for geothermal power plants are, therefore, small in comparison to solar power plants.

Comparing the amount of land required for a geothermal power facility with that required for a conventional fossil fuel or nuclear facility is more problematic. A geothermal power facility consists of multiple smaller individual plants, all of which are connected to the same thermal reservoir. In this sense, they occupy more land than conventional power plants, which generally produce as much or more energy as the entire collection of geothermal power plants but occupy a much smaller parcel of land. The problem with the comparison is that conventional power plants also depend upon mines, processing facilities, and an extensive transportation infrastructure to provide the fuel on which they depend and, in the case of coal and nuclear power plants, to remove the wastes generated by the production of electricity. By contrast, everything necessary to produce geothermal power is on the plot of land occupied by the plants themselves.

What may prove the most problematic aspect of geothermal power production is its dependence on large amounts of water. Probably the most famous example of poor water management occurred in New Zealand in a geothermal energy project at the Wairakei geothermal field. Water was extracted at high rates but not injected back into the field. The result was that the ground began to rapidly subside, in some places it collapsed at a rate of almost 1.6 feet (0.5 m) per year. Poor water management at the Geysers during the early 1980s permanently reduced the capacity of the system. Managing the water and creating a balance between what is withdrawn and what is injected may seem a simple bookkeeping problem in the sense that a plant operator must simply be sure to pump the same amount into the formation as was withdrawn, but the problem is more complex, as is made clear in the interview with John Farison in chapter 10.

Finally, enhanced geothermal systems (EGS) may make significant demands on the local water supply. If geothermal energy is to make a large contribution to energy production, it will be necessary to exploit many hot, dry rock formations. In these cases, significant amounts of surface water must be injected into the formation in order to fracture the rock and transport the thermal energy to the surface. Additionally, any water losses in the cycle must also be replaced with surface water. How much of a strain these plants will place on local water supplies will depend on the sites that are developed and the scale of development. But in the United States, where most of the best geothermal resources are located in the west, where water scarcity is more the norm than the exception, the development of these sites can be expected to further stress an already limited resource. Because commercial EGS facilities have not yet been built, no firm numbers are available.

Despite these challenges, it is no exaggeration to say that from an environmental point of view, there is currently no better power-generation technology than geothermal power. When measured by the impact they have on the environment per unit of power produced, they are almost ideal.

Government Policies and Geothermal Energy

In the United States, which has long been the world's leader in the development and production of geothermal energy, less than 1 percent of the nation's electrical needs are met with geothermal energy. While geothermal power plants deliver 20 percent of Iceland's electricity supply—most electricity in that nation is produced by hydroelectric plants—Iceland's unusually high percentage of geo-thermally produced power is more a reflection of its small economy than of its expansive network of geothermal power plants. (A few modestly sized geothermal power plants are sufficient to make a big impact on such a small economy.) Practically speaking, most large economies produce almost no power from geothermal energy—that is, if all geothermal power production were to suddenly cease, the shortfall could be quickly covered by other power producers, and most consumers would not even notice. In this sense, the contribution made by geothermal power to most economies is marginal. It

Taupo Volcanic Zone. New Zealand was one of the first nations to develop geothermal power. *(Sean McQueen)*

Danger
Dangerous ground
Keep to the boardwalk

will remain so as long as production is tied to the exploitation of formations of hot, permeable, water-saturated rock, which are too few in number to power a large geothermal industry.

Although the energy obtained from geothermal sources is relatively small, its promise is large. Interest in the technology was heightened in 2007 when a study entitled *The Future of Geothermal Energy, Impact of Enhanced Geothermal Systems (EGS) on the United States in the 21st Century* made the national news by asserting that "With reasonable R & D, EGS could provide 100 GWe or more of cost-competitive generating capacity in the next 50 years." (The abbreviation GWe means gigawatts of electric energy.) Such a prediction is very optimistic, and the paper is at least as concerned with selling the idea as analyzing it. Nevertheless, it remains true that beneath Earth's surface there is a tremendous supply of energy. It would be as unrealistic to dismiss the potential of geothermal energy as it would be to bet on the predictions made by the authors of the paper.

It is certainly true that if real change in geothermal energy production occurs, it will occur when enhanced geothermal systems become commercially viable. This technology, described in chapter 9, will enable power producers to mine the heat from many more formations than is currently possible. This chapter describes the main international effort to develop geothermal power and closes with a description of the state of geothermal power today.

THE INTERNATIONAL ENERGY AGENCY

International research into geothermal energy is carried out under the International Energy Agency Implementing Agreement for a Cooperative Programme on Geothermal Energy Research and Technology, which, not surprisingly, has an abbreviated title, the Geothermal Implementing Agreement (GIA).

The International Energy Agency (IEA), which is based in Paris, France, was established in 1974 in the wake of the oil crisis of 1973. The 1973 crisis began when some oil exporting countries unilaterally and sharply raised the price of oil and placed an oil embargo on the United States and the Netherlands for their support of Israel during the 1973 war between Israel and some of its neighbors. The result was severe economic disruption in many industrialized nations. In retrospect, it is hard to appreciate the shock of this first supply disruption. Industrialized nations had become accustomed to abundant and inexpensive supplies of oil, and they made use of oil in ways that made them even more dependent on those supplies than they are today. Many power plants, for example, burned oil to produce electricity, even base load electricity, a practice that is very unusual today. The IEA's founders envisioned the agency as an aggressive response to what was then called OPEC's "oil weapon," and the IEA was sometimes compared to the North Atlantic Treaty Organization (NATO). The IEA was supposed to be a collective response to threats to international energy supplies just as NATO was a collective Western response to military threats by the Soviet Union and its allies.

When the IEA was first created, there were 16 member states. (There are currently 27.) The document that established the organization emphasized the goal of creating systems that would mitigate the effects of any future supply disruptions. Member nations promised, for example, to maintain a supply of oil equaling at least 90 days of net oil imports and in the event of a disruption in supply to increase domestic oil production and suppress domestic demand. But other goals were also described, including the promotion of international collaborations to develop new energy technologies and alternative energy supplies. Devising ways to increase the energy security of member states remains a core part of the IEA's mission, but the goals that emphasized the development of alternative energy supplies and the value of using energy in environmentally responsible ways have received increased attention in more recent years. In particular, the GIA is now an important mechanism for furthering geothermal research and for sharing these results.

Originally envisaged to have a single five-year term, the GIA is now in its third five-year term, which is due to expire in 2012. The GIA document, which is available for inspection over the Internet, is long and formal and much of it treats matters of procedure as well as financing, insurance, and the ownership of patents arising from any research, but the appendices reflect the interests of those nations most concerned with geothermal energy. Each section lists the parties committed to pursuing a particular research topic, and each topic is divided into subtopics to facilitate the work. There are five main topics of interest: the environmental impacts of geothermal energy development, enhanced geothermal systems (EGS), deep geothermal resources, advanced geothermal drilling technology, and direct use of geothermal resources.

By way of example, Australia, the European Commission, Japan, Germany, Switzerland, and the United States are the participants in the EGS section. The following topics are to be addressed by the participants during their research into EGS during this term:

1. Enhanced Geothermal Systems Economic Models. As described in chapter 12, successful geothermal power plants are economic as well as technical concerns.

2. Application of Conventional Technology to Enhanced Geothermal Systems. This subtopic involves the identification and application of existing technologies to problems in geothermal energy production. Many of the technologies currently used for geothermal energy production were not developed with that application in mind. Most innovations in drilling technologies, for example, come from the petroleum business.

3. Data Acquisition and Processing. This involves the collection and analysis of information needed for the construction and operation of commercially viable geothermal power plants. (See the interview with John Farison in chapter 10 for an example of how this type of information is used at a conventional geothermal facility.)

4. Reservoir Evaluation. Understanding the characteristics of the geothermal reservoirs created by hydraulic fracturing is vital to predicting the performance of the resource. Geothermal energy projects are too expensive to simply try something in order to determine if it works.

5. Field Studies of EGS Reservoir Performance. The goal is to conduct field studies with an eye toward predicting reservoir characteristics and the maximum sustainable rates of energy recovery.

Probably the most high-profile IEA-GIA project is located at Soultz, France, where an international consortium is working to create an enhanced geothermal system. This may be the most ambitious and successful attempt yet to engineer a geothermal system. Engineers have successfully fractured hot, dry, impermeable rock to create a working thermal reservoir with a volume of about

(continued on page 176)

The Cost of Energy

One of the barriers faced by power producers interested in building more geothermal power plants is the high cost of building and operating those plants. Geothermal power plants, it is sometimes claimed, are just not economically competitive with more conventional power plants. But how does one determine the cost of any particular technology?

Members of the IEA have agreed not to adopt policies that intentionally distort energy prices. In particular, they agree not to undertake policies that, in order to meet social or industrial targets, result in artificially low prices for energy. The idea is for member governments to establish energy markets where the price of energy reflects the economic and environmental costs associated with its production and consumption. Artificially low prices, it is generally agreed, cause waste.

IEA members are quick to identify examples of such waste among non-IEA members. Iran, for example, is often criticized for subsidizing the cost of gasoline, which in 2007 was selling for about 34 cents per gallon (9 cents per liter) in that country. The effect of such a low price is to distort energy consumption patterns and to discourage efficiency. Iranians are famous for behaving as if gasoline were practically free, because for them it is. The Iranian government has even sponsored public information campaigns reminding motorists not to waste gasoline by spilling it onto the ground while fueling their cars.

Energy subsidies in the developed nations that comprise the membership of the IEA seem more difficult for IEA members to identify. In particular, there are many outside the organization who argue that the price of coal is held artificially low, especially in the United States. To be sure, in the United States, the low price of coal is one of its attractions, and electricity produced by coal-fired power plants is significantly less expensive than electricity produced by some other competing technologies, including geothermal power plants. Examining the nature of this price difference reveals a great deal about energy markets in the United States and elsewhere.

Currently, about half of all electricity produced in the United States comes from coal-fired power plants. But there are many who believe that the cost of coal in the United States (and many other IEA member states) no more reflects the true costs of producing and consuming coal than the price of gasoline in Iran reflects the true costs of its production and consumption. In the case of coal, some of those additional costs appear in the form of unchecked carbon dioxide emissions. Coal consumption, in particular, results in the emission of large amounts of carbon dioxide for each unit of heat produced. (Here, "large" means relative to any other major energy source.)

It is now known that the rising level of carbon dioxide in the atmosphere, which is caused, in part, by coal consumption, is an important contributor to climate change. The costs of climate change are borne by many people around the globe in the form of more droughts, more severe weather, and more crop failures, but these costs are not included in the price of coal. The benefits of inexpensive coal are shared by a relative few: Shareholders of coal-burning power plants distribute the profits among themselves, and their customers enjoy the benefits of inexpensive electricity. Meanwhile, those facing the highest costs associated with the large-scale consumption of coal often enjoy the fewest benefits.

Geothermal power production is, from an environmental viewpoint, a much cleaner technology than coal. It has fewer environmental costs. But historically it has been slow to develop, in part, because of the price differential between electricity produced by geothermal power plants and coal-fired power plants. This price difference would diminish—possibly disappear—depending on how much of the environmental costs associated with burning coal were included in the price of coal. If coal were priced more realistically, geothermal power—perhaps even some enhanced geothermal systems—would become more attractive to investors as coal became less attractive.

There are, however, two problems with establishing a more realistic price for coal. First, it is not clear how to compute such a price. (Knowing that a more realistic price exists is not enough to determine what the price should be.) Second, many individuals and businesses in IEA member countries would oppose increases in the price of coal, just as many Iranian individuals and businesses oppose increases in the price of gasoline, and for just the same reasons.

(continued from page 173)
one-half cubic mile (2 km³) at a depth of between 2.5–3 miles (4–5 km) and have produced hot water from this EGS system at rates that are close to commercial volumes.

The GIA is a good example of how international scientific research is conducted; it demonstrates the extent of international interest in geothermal energy production, and it facilitates research into those topics of most importance to the development of the technology. Finally, this document illustrates that there are many who believe that geothermal energy will have an important contribution to make in the 21st century.

GEOTHERMAL POWER PRODUCTION TODAY

What is the state of geothermal power today? Most geothermal production is centered along the Ring of Fire, a geologically active zone that encircles the Pacific Ocean, because in this zone geothermal energy is most easily accessed. What follows are brief descriptions of the world's five largest producers of geothermal electricity. Four of the producers are situated along the Ring of Fire.

The United States is the world's leader in electricity production from geothermal energy with more than two GW of installed capacity. (For purposes of comparison, an average-sized nuclear plant will produce about one GW of power, and about 100 nuclear power plants are in operation today in the United States.) Among the states, California produces more electricity from geothermal power than all the other states combined. States with more modest levels of geothermal power production include Nevada, Utah, and Hawaii. The relatively large scale at which geothermal power production is deployed in the United States is only partially a reflection of the nation's relatively large supply of easily accessed geothermal energy. The industry is also supported by generous government subsidies, including modest amounts of government-funded research and production tax credits, which grant geothermal producers tax credits for each kilowatt-hour of electricity produced. (A kilowatt-

California State Capitol. California, which is located along the Pacific Ring of Fire, is a world leader in geothermal power. *(Kevin D. Korenthal)*

hour, abbreviated kWh, is way of measuring energy production. A kilowatt-hour is produced, for example, when a plant produces one kilowatt of electricity continuously for one hour or when it produces two kilowatts of electricity continuously for one-half hour.) Despite this, geothermal energy supplies less than 1 percent of the nation's electricity needs. In addition to electricity production, direct-use applications make a small contribution to the nation's energy supply, and more geothermal heat pumps are used in the United States than in any other nation.

The nation with the second highest level of geothermal electricity production is the Philippines. It currently has approximately 1.8 GW of installed geothermal power capacity, and is seeking to build another gigawatt of capacity by 2016. The situation in the Philippines is currently in flux as the government restructures the electricity sector. If the Philippines is successful in building this

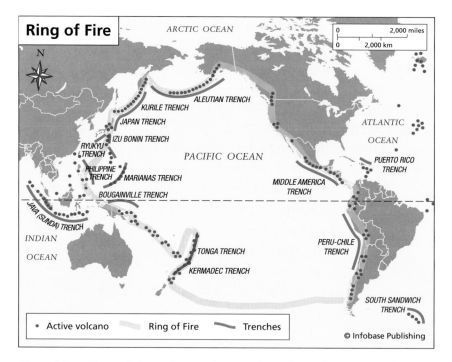

Ring of Fire. Most of the major producers of geothermal energy are located along the Ring of Fire.

additional capacity, it may surpass the United States as the world's largest geothermal power producer. Geothermal energy supplies about 7 percent of that nation's electricity needs.

Mexico has an installed capacity of almost one gigawatt of electricity of geothermal power. There is little government support for the technology at the present time, but the geothermal power sector remains reasonably vibrant. Plans are being pursued to update older facilities, install new units at geothermal fields already in operation, and develop up to four new fields. Roughly 3 percent of that nation's electricity comes from geothermal power sources.

Indonesia has an installed geothermal capacity that is almost equal to that of Mexico—about 0.8 GW. Geothermal development in Indonesia suffered during and after the 1997–98 economic difficulties that affected Indonesia and a number of its Asian neighbors. Political instability has further compounded problems in development of the resource. A number of important geothermal projects have been postponed or canceled. Today, the prices paid for electricity by the government are too low to attract investors. Although the regulations are currently under revision with the goal of reducing development costs and attracting additional geothermal investment in this land of enormous geothermal resources, all parties acknowledge that it will now take years for investors to return to the Indonesian energy markets. Currently, geothermal energy supplies about 4 percent of the Indonesia's electricity needs.

The fifth largest producer of geothermal electricity is also the world's oldest. Geothermal power began in 1904 in Larderello, Italy, the site of a large dry steam field. There are about 0.8 GW of installed capacity in Italy. The industry is buoyed by generous subsidies, and new facilities are in various stages of planning and construction. Today, geothermal energy supplies approximately 2 percent of Italy's electricity needs.

These figures make it clear that geothermal energy remains on the margins of the energy sector even in those nations with the largest geothermal power capacities. Whether geothermal energy becomes a major contributor to the world's supply of electricity will depend on the solution of certain technical problems associated with the development of EGS as well as the evolving costs of competing forms of energy. Earth's interior holds an inexhaustible supply of thermal energy, and virtually all of it is untapped. Geothermal technology offers the promise of harnessing that source of energy to provide a continuous, highly reliable source of power produced with minimal environmental impacts, and the necessary energy is only a few (vertical) miles away. The future of geothermal power seems bright.

Chronology

ca. 300 B.C.E.	Greeks use lenses to concentrate sunlight
ca. 200 B.C.E.	Greeks use mirrors to concentrate sunlight
212 B.C.E.	Mirrors may have been used to concentrate sunlight for use as a weapon during the siege by the Romans of the Greek city-state of Syracuse
1839	French scientist Alexandre-Edmond Becquerel produces an electric current by shining light on certain metals immersed in a chemical solution
1880	Alexander Graham Bell and Charles Sumner Tainter invent the photophone
1883	American inventor Charles Fritts publishes a paper describing his invention, the first solar cell
1891	First commercial solar water heater patented by Clarence Kemp of Baltimore, Maryland
1892	World's first geothermal heating district begins operation in Boise, Idaho
1904	First electrical generator powered with geothermal energy begins operation at Larderello, Italy
1905	German-born American physicist Albert Einstein publishes his paper on the photoelectric effect
1948	First geothermal heat pump goes into operation at home of Carl Nielsen, physics professor, Ohio State University

1954 Crystalline silicon-based photovoltaic technology developed at Bell Laboratories. These cells had a conversion efficiency of about 4 percent

1957 Soviet Union launches first Sputnik spacecraft, initiating a race between the Soviet Union and the United States to develop better space hardware, including better ways of powering space systems

1958 *Sputnik 3, Vanguard 1, Vanguard 2,* and *Explorer 3* all launched with PV arrays

First generating station begins operation at Wairakei, New Zealand; it is also the first system in the world to utilize hot water rather than dry steam

1960 First commercial-scale generating units begin operation at the Geysers

1962 Telstar, first telecommunications satellite, is launched. It is powered by PV arrays

1964 NASA launches *Nimbus 1,* a satellite powered by a 470-watt PV array

1965 American engineer Peter Glaser conceives of space-based solar power generating station

1969 World's first solar furnace constructed at Odeillo, France

1970 First enhanced geothermal test facility—later built at Fenton Hill, New Mexico—is proposed by a team at Los Alamos National Laboratory

1973 OPEC sharply increases prices. Some OPEC members place an embargo on shipments of oil to the United States and the Netherlands

1974 First borehole to test feasibility of enhanced geothermal systems is drilled at Fenton Hill test site

1978 U.S. Congress passes Public Utilities Regulatory Policies Act (PURPA) and Energy Tax Act

1979 Political instability in Iran results in second energy crisis of the 1970s. Oil prices increase sharply again

1981 First demonstration binary geothermal power plant begins operation in California's Imperial Valley

1982 A one-megawatt PV power station begins operation in Hisperia, California, and Solar One, a 10-megawatt power tower, begins operation near Barstow, California

1987 Work begins on the experimental enhanced geothermal systems facility at Soultz, France

1989 Power output at the Geysers, the world's largest geothermal electricity generating complex, peaks

1994 Fenton Hill research program ended

1996 The Federal Energy Regulatory Commission issues orders 888 and 889, thereby restructuring the electricity markets in the United States

Solar Two, using an innovative thermal storage system, begins operation in Daggett, California

2001 The National Space Development Agency of Japan begins work on space-based solar power system that will beam energy back to Earth's surface

2004 Germany adopts the Renewable Energy Sources Act

2005 U.S. Congress passes the Energy Policy Act

2007 The Defense Advanced Research Projects Agency announces plans to develop PV cells that convert in excess of 50 percent of incident sunlight into electricity

Prince Hassan bin Talal of Jordan presents a paper entitled "Clean Power from Deserts—The DESERTEC Concept for Energy, Water, and Climate Security" to members of the European Parliament, describing a proposal in which Europe purchases substantial amounts of power from huge concentrating solar power (CSP) systems deployed throughout the Middle East and North Africa (MENA), and in return MENA receives freshwater, which is produced as a coproduct with the electricity, as well as income and electricity

2008   A dish/engine system at Albuquerque, New Mexico, achieves a 31.25 percent solar-to-electricity conversion efficiency, a world's record for CSP systems

# List of Acronyms

Btu	British Thermal Unit
CSP	Concentrating Solar Power
EGS	Enhanced Geothermal Systems
GIA	International Energy Agency Implementing Agreement for a Cooperative Programme on Geothermal Energy Research and Technology (Geothermal Implementing Agreement)
IEA	International Energy Agency
kW	kilowatt
kWh	kilowatt-hour
MW	megawatt
MWh	megawatt-hour
nm	nanometer
PV	photovoltaic

# Glossary

base load power   the minimum amount of power required by an electric market over a given period of time

binary geothermal power   a technology that uses water to transfer heat from a geological formation to a secondary working fluid for the purpose of generating electricity

British thermal unit (Btu)   unit of energy equal to 1,056 joules

capacity factor   the ratio of the amount of power produced by a generating unit over a representative period of time to the amount of power produced by that unit when operated continuously at full power over the same time period

concentrating solar power system (CSP)   electricity-generating station that uses focused sunlight as a source of thermal energy

demand management   a system for modifying the pattern of consumer electricity usage, largely by pricing electricity according to the time that it is used as well as by the amount consumed

direct use   technologies that involve the use of geothermal energy without first converting that energy into electricity

dish/engine system   an electricity-generating unit that depends on a large parabolic mirror (dish) to concentrate sunlight. The resulting heat drives a high-efficiency engine

distributed generation   a system for supplying power that depends on numerous small-scale generating units, each of which is typically located near to the point of use

electromagnetic wave   energy moving at the speed of light and consisting of regular pulsations in the intensity of electric and magnetic energy; e.g., light waves and radio waves

enhanced geothermal systems (EGS)   engineered geothermal systems designed to exploit formations of hot, dry rock, hot impermeable rock, or hot, dry, and impermeable rock

geothermal heat pump   a machine that transfers thermal energy between the interior of a building and the ground outside of it

green architecture   often used to describe so-called sustainable building designs; here used more narrowly to describe architecture that seeks to make optimal use of local renewable energy supplies

heat engine   a machine for the conversion of thermal energy into mechanical or electrical energy

heat exchanger   a device that permits the transfer of heat but not mass between two fluids

hot spot   a stationary spot beneath a tectonic plate through which magma rises. The result is a sequence of volcanoes as the tectonic plate moves across the "fountain" of lava.

infrared waves   electromagnetic waves with wavelengths between about one mm and 700 nm

lithosphere   solid outer layer of planet Earth consisting of the crust and the uppermost part of the mantle

magma chamber   highly porous formation of solid rock, the pores of which are filled with molten rock

megawatt   one million watts

megawatt-hour   unit of energy equivalent to the continuous output of one megawatt for one hour

nanometer   one-billionth of a meter

net metering   the practice of issuing a utility bill (or credit) equal to the amount of electricity consumed minus the amount of electricity injected back into the grid during a billing period. A negative amount represents a credit

parabolic trough   a solar concentrator in the form of a long trough with a cross section in the shape of a parabola; a solar generating station that uses parabolic troughs to concentrate sunlight

passive solar   building designs that use the Sun for heating, cooling, ventilation, or lighting without relying on photovoltaic or concentrating solar power conversion technologies

peak load   here used for that portion of the electrical demand that exceeds base load power demand

phase change   a change in the state of matter, from solid to liquid, for example, or from liquid to gas

photovoltaic (PV)   technology used to convert solar energy directly into electrical energy

plane of the ecliptic   the plane that contains Earth's orbital path about the Sun

plate tectonics   the geological theory that asserts that Earth's lithosphere is fragmented into a number of plates that are in slow but continuous motion

power tower   a power-generating station that uses many large movable mirrors called heliostats to focus the Sun's light on a heat exchanger, located at the top of a tower placed above the field of mirrors

semiconductor   a solid material with electrical properties that are midway between electrical conductors and electrical insulators. Photovoltaic technology makes essential use of semiconductors

solar constant   a quantity used to represent the average amount of solar energy received at a point above Earth's atmosphere at a distance from the Sun equal to the average Earth-Sun distance, often given as 1,368 watts per square meter

specific heat   the thermal energy necessary to raise the temperature of a unit mass of material by one degree

ultraviolet waves   electromagnetic waves with wavelengths between 400 nm and 1 nm

watt   a unit of power equal to one joule per second

wavelength   the distance between two successive peaks (or troughs) of a wave

working fluid   a fluid (liquid or gaseous) used to transfer heat

Further Resources

Currently, solar and geothermal energy are niche technologies in the sense that they make relatively small contributions to the power market. But their small contributions belie their large potential. To understand why they have remained such small contributors to the power markets, one must, of course, understand the physics and economics of these technologies, but one should also understand why replacing one energy infrastructure with another is so difficult even if the new technology seems clearly better than the old. These issues are addressed in the following books and articles.

BOOKS

Duffield, Wendell A., and John H. Sass. *Geothermal Energy: Clean Power from the Earth's Heat.* Menlo Park, Calif.: U.S. Geological Survey Information Services, 2003. A brief, well-written introduction to the subject.

Erickson, Jon. *Plate Tectonics: Unraveling the Mysteries of the Earth.* New York: Facts On File, 2001. To understand the distribution of geothermal resources, it is necessary to understand the basic principles of plate tectonics. This is a highly readable introduction.

Gupta, Harsh K. *Geothermal Energy: An Alternative Resource for the 21st Century.* Boston: Elsevier, 2007. A more technical and more thorough account of the concepts and technologies involved in the exploitation of geothermal energy.

Houghton, John. *Global Warming: The Complete Briefing.* 3d ed. Cambridge: Cambridge University Press, 2004. Global warming is the topic that drives much of the interest in so-called alternative forms of energy. *Global Warming* offers a thoughtful and reasonably thorough description of what is presently known about the phenomenon of human-induced climate change.

Kryza, Frank T. *The Power of Light: The Epic Story of Man's Quest to Harness the Sun.* New York: McGraw-Hill, 2003. The 19th- and early 20th-century history of concentrating solar power systems is the subject matter of this unique, entertaining, and very informative book. Highly recommended.

Messenger, Roger A., and Jerry Ventre. *Photovoltaic Systems Engineering.* 2nd ed. New York: CRC Press, 2004. This book provides a thorough discussion of photovoltaics beginning at an elementary level and ending with some fairly advanced technical discussions.

National Research Council, Committee for the Assessment of NASA's Space Solar Power Investment Strategy. *Laying the Foundation for Space Solar Power: An Assessment of NASA's Space Solar Power Investment Strategy.* Washington, D.C.: National Academies Press, 2001. It is also available online in a very user-unfriendly format. URL: http://www.nap.edu/openbook. php?record_id=10202&page=R1. This is a brief description of the proposal to build a solar-power generating station in outer space to provide power for terrestrial applications, as well as a review of NASA's efforts to this effect.

National Research Council, Committee on Nuclear and Alternative Energy Systems. *Energy in Transition, 1985–2010: Final Report of the Committee on Nuclear and Alternative Energy Systems.* San Francisco: W.H. Freeman, 1980. There are currently many books on the market that describe the rapid introduction of solar energy technology as inevitable; they downplay the difficulties involved. This almost 700-page book was written by some of the best engineers and scientists in United States at the time, and they got almost everything wrong. For example, only four pages of this tome are devoted to wind energy, but whole chapters are devoted to breeder reactors, controlled nuclear fusion, and solar energy. It turns out that in the United States wind energy now produces more power than all three of the other sources combined, an important reminder that although a solar energy–based economy may (or may not) be inevitable, it will certainly not be easy to create. Also available online. URL: http://books.nap.edu/openbook.php?record_id=11771&page=R1.

Pool, Robert. *Beyond Engineering: How Society Shapes Technology.* New York: Oxford University Press, 1997. An interesting book that is concerned with the general problem of how society affects technology. While not specifically about solar or geothermal energy, it is an important discussion about how technology evolves.

Tabak, John. *Nuclear Power.* New York: Facts On File, 2009. This volume contains more detailed discussions of heat engines and the concept of thermal efficiency.

INTERNET RESOURCES

Bentley, Molly. "Guns and Sunshades to Rescue Climate." Available online. URL: http://news.bbc.co.uk/2/hi/science/nature/4762720.stm. Accessed on February 25, 2008. One of the most compelling arguments for the use of solar and geothermal energy is that their use does not alter the climate. Some argue that the only realistic way to avoid climate change, however, is to alter the amount of sunlight reaching Earth. Here is a report on this idea.

California Public Utilities Commission. "San Diego Gas & Electric's Proposed Sunrise Powerlink Transmission Project." Available online. URL: http://www.cpuc.ca.gov/puc/hottopics/1energy/a0512014.htm. Accessed on February 25, 2008. This page is maintained by the California Public Utilities Commission and contains important information on whether a high voltage transmission line should be constructed through public land to bring power from the Imperial Valley to customers in the San Diego metropolitan area.

IEA Geothermal Energy. Available online. URL: http://www.iea-gia.org/. Accessed on February 25, 2008. The home page for the International Energy Agency's geothermal program. It contains the text of the geothermal implementing agreement and important information about its activities.

Illinois Institute of Technology. *Design of a 20-Megawatt Geothermal Power Plant.* Available online. URL: http://www.iit.edu/~ipro348s07/process_flow.html. This site gives a nice overview of the workings of a binary geothermal power plant and provides an animated illustration of the process.

National Renewable Energy Laboratory. "Solar Maps." Available online. URL: http://www.nrel.gov/gis/solar.html. Accessed on February 25, 2008. This is an online atlas that details the distribution of solar energy in the United States by technology type, month, and other characteristics.

Osborn, Julie, and Cornelia Kawann. *Reliability of the U.S. Electricity System: Recent Trends and Current Issues.* Available online. URL: http://eetd.lbl.gov/ea/EMP/reports/47043.pdf. Accessed on February 25, 2008. This report, prepared for the U.S. Department of Energy, is a very general discussion of system reliability. It is a nice introduction to this very important topic, which gains in importance as more intermittent suppliers of power are integrated into the system.

Sanner, Burkhard. "Shallow Geothermal Energy." Available online. URL: http://geoheat.oit.edu/bulletin/bull22-2/art4.pdf. Accessed on February 25, 2008. A well-written introduction to the theory and technology of geothermal heat pumps.

Tester, Jefferson W., et al. *The Future of Geothermal Energy: Impact of Enhanced Geothermal Systems (EGS) on the United States in the 21st Century.* Available online. URL: http://geothermal.inel. gov/publications/future_of_geothermal_energy.pdf. Accessed on February 25, 2008. This highly optimistic paper made a big impact when it was first published in 2006, describing the enormous potential of enhanced geothermal systems, although the first commercial EGS power plant had yet to be built.

United States Department of Energy, Office of Energy Efficiency and Renewable Energy. *Geothermal Basics Overview.* Available online. URL: http://www1.eere.energy.gov/geothermal/ geothermal_basics.html. Accessed on February 25, 2008. This site and its links provide a very basic but well-written overview of geothermal energy technology, history, and current research.

United States Department of Energy, Office of Energy Efficiency and Renewable Energy. *Solar Energy Technologies Program.* Available online. URL: http://www1.eere.energy.gov/solar/. Accessed on February 25, 2008. This site and its links provide a very basic but well-written overview of all of the basic solar energy technologies, their history, and current research.

Index

Note: *Italic* page numbers indicate illustrations and diagrams; page numbers followed by *c* indicate chronology entries.